Improving the Air Force Scientific Discovery Mission

Leveraging Best Practices in Basic Research Management

A WORKSHOP REPORT

Committee on Improving the Air Force Scientific Discovery Mission:
Leveraging Best Practices in Basic Research Management:
A Workshop

Air Force Studies Board

Division on Engineering and Physical Sciences

The National Academies of
SCIENCES · ENGINEERING · MEDICINE

THE NATIONAL ACADEMIES PRESS
Washington, DC
www.nap.edu

THE NATIONAL ACADEMIES PRESS   500 Fifth Street, NW   Washington, DC 20001

This is a report of work supported by Grant FA9550-14-1-0127 between the U.S. Air Force and the National Academy of Sciences. Any opinions, findings, or conclusions expressed in this publication are those of the author(s) and do not necessarily reflect the view of the organizations or agencies that provided support for the project.

International Standard Book Number-13:   978-0-309-37830-7
International Standard Book Number-10:   0-309-37830-3

Limited copies of this report are available from:

Additional copies are available from:

Air Force Studies Board
National Research Council
500 Fifth Street, NW
Washington, DC 20001
(202) 334-3111

The National Academies Press
500 Fifth Street, NW
Keck 360
Washington, DC 20001
(800) 624-6242 or (202) 334-3313
http://www.nap.edu

Copyright 2015 by the National Academy of Sciences. All rights reserved.

Printed in the United States of America

Suggested citation: National Academies of Sciences, Engineering, and Medicine. 2015. *Improving the Air Force Scientific Discovery Mission: Leveraging Best Practices in Basic Research Management: A Workshop Report.* Washington, DC: The National Academies Press.

The National Academies of SCIENCES · ENGINEERING · MEDICINE

The **National Academy of Sciences** was established in 1863 by an Act of Congress, signed by President Lincoln, as a private, nongovernmental institution to advise the nation on issues related to science and technology. Members are elected by their peers for outstanding contributions to research. Dr. Ralph J. Cicerone is president.

The **National Academy of Engineering** was established in 1964 under the charter of the National Academy of Sciences to bring the practices of engineering to advising the nation. Members are elected by their peers for extraordinary contributions to engineering. Dr. C. D. Mote, Jr., is president.

The **National Academy of Medicine** (formerly the Institute of Medicine) was established in 1970 under the charter of the National Academy of Sciences to advise the nation on medical and health issues. Members are elected by their peers for distinguished contributions to medicine and health. Dr. Victor J. Dzau is president.

The three Academies work together as the **National Academies of Sciences, Engineering, and Medicine** to provide independent, objective analysis and advice to the nation and conduct other activities to solve complex problems and inform public policy decisions. The Academies also encourage education and research, recognize outstanding contributions to knowledge, and increase public understanding in matters of science, engineering, and medicine.

Learn more about the National Academies of Sciences, Engineering, and Medicine at **www.national-academies.org**.

COMMITTEE ON IMPROVING THE AIR FORCE SCIENTIFIC DISCOVERY MISSION: LEVERAGING BEST PRACTICES IN BASIC RESEARCH MANAGEMENT: A WORKSHOP

LARRY D. WELCH, Institute for Defense Anlyses, *Chair*
RITA R. COLWELL, University of Maryland, College Park
BLAISE J. DURANTE, U.S. Air Force (retired)
MELISSA FLAGG, John D. and Catherine T. MacArthur Foundation
BRENDAN B. GODFREY, University of Maryland, College Park
ZACHARY J. LEMNIOS, IBM Research Division
WILLIAM L. MELVIN, Georgia Tech Research Institute
PARVIZ MOIN, Stanford University
ROBIE I. SAMANTA ROY, Lockheed Martin Corporation
SUBHASH C. SINGHAL, Pacific Northwest National Laboratory

Staff

CARTER W. FORD, Program Officer
MARGUERITE E. SCHNEIDER, Administrative Coordinator
DIONNA C. ALI, Research Assistant

AIR FORCE STUDIES BOARD

DOUGLAS M. FRASER, Doug Fraser, LLC, *Chair*
DONALD C. FRASER, Charles Stark Draper Laboratory (retired), *Vice Chair*
BRIAN A. ARNOLD, U.S. Air Force (retired), Peachtree City, Georgia
ALLISON ASTORINO-COURTOIS, National Security Innovations, Inc.
TED F. BOWLDS, The Spectrum Group
STEVEN R.J. BRUECK, University of New Mexico
FRANK J. CAPPUCCIO, Cappuccio and Associates, LLC
BLAISE J. DURANTE, U.S. Air Force (retired)
MELISSA L. FLAGG, John D. and Catherine T. MacArthur Foundation
BRENDAN B. GODFREY, University of Maryland, College Park
MICHAEL A. HAMEL, Lockheed Martin Space Systems Company
DANIEL E. HASTINGS, Massachusetts Institute of Technology
RAYMOND E. JOHNS, JR., Flight Safety International
ROBERT H. LATIFF, R. Latiff Associates
NANCY G. LEVESON, Massachusetts Institute of Technology
MARK J. LEWIS, Institute for Defense Analyses, Science and Technology Policy Institute
ALEX MILLER, University of Tennessee
OZDEN OCHOA, Texas A&M University
RICHARD V. REYNOLDS, The VanFleet Group, LLC
STARNES E. WALKER, University of Delaware
DEBORAH WESTPHAL, Toffler Associates
DAVID A. WHELAN, Boeing Defense, Space, and Security
REBECCA WINSTON, Winston Strategic Management Consulting
MICHAEL I. YARYMOVYCH, Sarasota Space Associates

Staff

JOAN FULLER, Director
ALAN H. SHAW, Deputy Director
DIONNA C. ALI, Research Assistant
CARTER W. FORD, Program Officer
CHRIS JONES, Financial Manager
ANDREW J. KREEGER, Program Officer
MARGUERITE E. SCHNEIDER, Administrative Coordinator

Acknowledgment of Reviewers

This report has been reviewed in draft form by individuals chosen for their diverse perspectives and technical expertise, in accordance with procedures approved by the Report Review Committee. The purpose of this independent review is to provide candid and critical comments that will assist the institution in making its published report as sound as possible and to ensure that the report meets institutional standards for objectivity, evidence, and responsiveness to the study charge. The review comments and draft manuscript remain confidential to protect the integrity of the deliberative process. We wish to thank the following individuals for their review of this report:

Steven R.J. Brueck, University of New Mexico,
Donald R. Erbschloe, Air Mobility Command, U.S. Air Force,
Janet S. Fender, Air Combat Command, U.S. Air Force,
Daniel E. Hastings, Massachusetts Institute of Technology,
Chad A. Mirkin, Northwestern University, and
Lyle H. Schwartz, U.S. Air Force (retired).

Although the reviewers listed above have provided many constructive comments and suggestions, they were not asked to endorse the views presented at the workshop, nor did they see the final draft of the workshop report before its release. The review of this workshop report was overseen by Robert J. Hermann, U.S. Air Force (retired), who was responsible for making certain that an independent examination of this workshop report was carried out in accordance with institutional procedures and that all review comments were carefully considered. Responsibility for the final content of this report rests entirely with the committee and the institution.

Contents

OVERVIEW 1

1 CONTEXT FOR THE WORKSHOP 4
  Organization of the Workshop, 6
  Management Practices Considered in the Workshop, 6
  Recommendations from Previous Studies, 7

2 WORKSHOP PARTICIPANT OBSERVATIONS AND KEY THEMES 9
  AFOSR's Mission, 9
  AFOSR's Mission Leadership, 11
  AFOSR's Mission Force, 12
  AFOSR's Mission Support, 14
  Transitioning Mission Success, 15
  Perspectives of a Former U.S. Air Force Chief of Staff, 16

3 SYNOPSIS OF WORKSHOP SESSIONS 18
  Session 1, April 27-28, 2015, 18
  Session 2, May 27-28, 2015, 29

APPENDIXES

A Terms of Reference 37
B Biographical Sketches of Committee Members 38
C Workshop Sessions and Speakers 42
D Excerpt from the Carl Levin National Defense Authorization Act for Fiscal Year 2015 45

Acronyms

AFOSR	Air Force Office of Scientific Research
AFRL	Air Force Research Laboratory
ASD(R&E)	Office of the Assistant Secretary of Defense for Research and Engineering
DARPA	Defense Advanced Research Projects Agency
DDR&E	Office of the Director of Defense Research and Engineering
DoD	Department of Defense
DoE	Department of Energy
DTIC	Defense Technical Information Service
FFRDC	federally funded research and development center
IPA	Intergovernmental Personnel Act
IRB	Institutional Review Board
ISR	intelligence, surveillance, and reconnaissance
MAJCOM	major command
NRC	National Research Council
NSF	National Science Foundation
ONR	Office of Naval Research
OSD	Office of the Secretary of Defense
OSTP	Office of Science and Technology Policy
S&E	scientist and engineer
S&T	science and technology
SAF/AQR	Deputy Assistant Secretary of the Air Force for Science, Technology, and Engineering
SES	Senior Executive Service Position

ST	scientific or professional position
STTR	Small Business Technology Transfer
UARC	University Affiliated Research Center
UCC	Unified Combat Command
USAFA	U.S. Air Force Academy
USD(AT&L)	Under Secretary of Defense for Acquisition, Technology, and Logistics

Overview

The Air Force Office of Scientific Research (AFOSR) was established in October 1951 to manage the Air Force's basic research portfolio.[1] AFOSR is one of nine directorates of the Air Force Research Laboratory (AFRL) and manages the $450.1 million U.S. Air Force basic research budget.[2] All but about $114.8 million of this budget is awarded as research grants to universities.[3] The remaining funds are directed towards overhead and funding provided to government offices, including the technical directorates at AFRL. In addition to stimulating research discoveries, these funds directly and indirectly support science, technology, engineering, and mathematics graduate education and strengthen ties between the Air Force and academia.

In 2014, the Deputy Assistant Secretary of the Air Force for Science, Technology, and Engineering and the director of AFOSR requested that the Air Force Studies Board of the National Academies of Sciences, Engineering, and Medicine undertake two 2-day workshops to (1) explore the unique drivers associated with management of a 6.1 basic research portfolio in the Department of Defense (DoD) and establish the current AFOSR baseline business practices across all its functional offices; (2) review Army, Navy, and Office of the Secretary of Defense (OSD) practices for management of basic research that could be benchmarked by AFOSR for incorporation to enhance its own practices; and (3) facilitate a discussion with AFOSR stakeholders (Air Force, OSD, the Office of Science and Technology Policy, and Congress) as to current and future practices that may further the effective and efficient management of 6.1 basic research on behalf of the Air Force. At the time the workshop was requested, there were discussions within and without the Air Force to relocate AFOSR headquarters from the Washington, D.C., area to Wright-Patterson Air Force Base in Ohio. The issue has since been resolved to keep AFOSR in Washington, D.C. Thus, the topic was not relevant to the scope of the workshop.

Across the two workshop sessions, several common messages appeared to resonate with multiple participants stemming from various presentations and resulting dialog among the participants. These common messages fall under the categories of "AFOSR's Mission;" AFOSR's Mission Leadership;" "AFOSR's Mission Force;" "AFOSR's Mission Support;" and "Transitioning Mission Success" and are summarized below. Chapter 2 of this report provides additional discussion for each topic. Finally, the report summarizes the views expressed by individual workshop

---

[1] Air Force Office of Scientific Research, *Air Force Office of Scientific Research 1951-Present: Turning Scientific Discovery Into Air Force Opportunity*, http://www.wpafb.af.mil/library/factsheets/factsheet.asp?id=8976, accessed November 21, 2014.

[2] Mark Amundson, Air Force Office of Scientific Research, personal communication to Carter Ford on June 18, 2015. Note: The figures reflect Fiscal Year 2014.

[3] Ibid.

participants. While the committee is responsible for the overall quality and accuracy of the report as a record of what transpired at the workshops, the views contained in this section and in the rest of the report are not necessarily those of all workshop participants, the committee, or the Academies.

1. *AFOSR's Mission.* Several participants in both workshop sessions noted that AFOSR is a unique organization that fulfills a mission important to national security—connecting the basic research community to near- and far-term national security needs. Moreover, many of these participants highlighted that AFOSR helps to support growth in depth and breadth of the best and brightest, who develop from individual investigators to leaders of their fields, while at the same time maintaining strong ties between the future leaders and the Air Force. It was noted by some workshop participants that AFOSR international offices are especially critical to building ties with leading researchers throughout the world, obtaining current knowledge about foreign scientific advances, and building international good will. Colocation of AFOSR with the Office of Naval Research, the Defense Advanced Research Projects Agency, the National Science Foundation, the Department of Energy's Office of Science, the Office of the Assistant Secretary of Defense for Research and Engineering, and other federal funding agencies in the Washington, D.C., metropolitan area is highly beneficial, some participants stated.

2. *AFOSR's Mission Leadership.* Past and current directors of AFOSR commented at both workshop sessions on the essential need for stable, qualified, committed leadership at the top of AFOSR for effective mission performance. These leaders stressed that the AFOSR director and other senior leaders should be expected to occupy their positions for at least 4 years. In comparison with basic research organizations in other Services, AFOSR has a disproportionately small number of Senior Executive Service (SES) positions and other senior civilian positions for its budget. As noted by the past AFOSR directors, some AFOSR support organizations are not responsible to the AFOSR director, and this reduces the ability of the organization to accomplish its mission. Many of these directors noted that DoD treats basic research funds as though they were 1-year funds rather than 2-year funds, which, they argue, is inappropriate for basic research and harmful to AFOSR performance. Budget instability also is an issue.

3. *AFOSR's Mission Force.* As epitomized in the first workshop session, the mission execution force for the work of the AFSOR is its cadre of program officers. Multiple workshop participants noted that the efficiency and effectiveness of AFSOR in fulfilling its mission rests on the effectiveness and efficiency of this mission execution team. In that context, these workshop participants stated that the highest-quality program officers—with high levels of domain technical competence and the leadership and management experience to lead basic research—are essential to the success of AFOSR. Unfortunately, AFOSR is understaffed from its optimum level of program officers by about 25 percent, and important research efforts are being reduced due to vacancies. Multiple speakers at both workshop sessions cited the importance of recruiting and retaining the right number of well-qualified program officers, including through the Intergovernmental Personnel Act.

4. *AFOSR's Mission Support.* Many participants at both workshop sessions stated that practices and processes that facilitate and support discovering and leveraging the best basic research wherever it occurs are essential. Others stated that establishing sustained relationships that produce relevant proposals, funded grants, and other interactions with the best and brightest researchers in the national security arena are also essential. However, some participants asserted that limitations on professional conference and workshop attendance, and on access to journals and big data resources, inhibit AFOSR researchers from keeping pace with scientific and engineering advances, both foreign and domestic. Excellent electronic workflow and other internal information technology business processes have only partly offset the loss of assistant program officers and other support personnel. Program officers at the first workshop session repeatedly raised the issue of time spent doing activities that could be performed by support personnel is time lost from performing the mission.

5. *Transitioning Mission Success.* Technology transition from AFOSR relies primarily on personal relationships established by program officers. Although many participants noted that this ad hoc procedure often is effective, they also argued that orderly processes for transitioning basic research discoveries to other parts of the Air Force should be established and institutionalized. Additionally, multiple participants believed that AFOSR should document better the many important science and engineering advances it has sponsored.

ORGANIZATION OF THE REPORT

Chapter 1 provides the context for the workshop. Chapter 2 provides workshop participant observations grouped under the following headings: AFOSR's Mission; AFOSR's Mission Leadership; AFOSR's Mission Force; AFOSR's Mission Support; and Transitioning Mission Success. Chapter 2 concludes with thoughts provided by General Larry Welch (USAF, Ret.), a former chief of staff of the U.S. Air Force. Chapter 3 then provides synopses of the two workshop sessions.

1

Context for the Workshop

Throughout its history, the U.S. Air Force has relied on highly skilled scientists and technologists to manage the discovery, development, and integration of science and technology to advance the fielding of advanced Air Force weapon systems. Within the Air Force, which has historically placed a premium on scientific discovery, the Air Force Office of Scientific Research (AFOSR) has responsibility for overseeing investments in basic research for Air Force applications.[1]

> AFOSR continues to expand the horizon of scientific knowledge through its leadership and management of the Air Force's basic research program. As a vital component of the Air Force Research Laboratory (AFRL), AFOSR's mission is to support Air Force goals of control and maximum utilization of air, space, and cyberspace.[2]

Over the years, AFOSR has played a large role in developing and nurturing talent in the nation. This starts with the graduate students that AFOSR funds at the universities and then also includes principal investigators who have been nurtured as assistant professors and whose career has developed as a result. Both the Air Force and Department of Defense (DoD) have recognized the challenges in managing defense-focused scientific research in academia due to the uniqueness of educational institutions, Presidential and DoD basic research policy, and specialized skills and business practices required by basic research managers to ensure effectiveness and efficiency of the investment. For purposes of this report, DoD basic research is defined as follows:

> Basic research is systematic study directed toward greater knowledge or understanding of the fundamental aspects of phenomena and of observable facts without specific applications towards processes or products in mind. It includes all scientific study and experimentation directed toward increasing fundamental knowledge and understanding in those fields of the physical, engineering, environmental, and life sciences related to long-term national security needs. It is farsighted high payoff research that provides the basis for technological progress. Basic research may lead to: (a) subsequent applied research and advanced technology developments in Defense-related technologies, and (b) new and improved military functional capabilities in areas such as communications, detection, tracking, surveillance,

---

[1] Air Force Research Laboratory, *Air Force Office of Scientific Research 2014 Technical Strategic Plan*, 2014, https://community.apan.org/afosr/spring_review_2014/m/spring_review_2014_non_presentation_files/132008.aspx.

[2] Wright-Patterson Air Force Base, "AFOSR: About – Mission," posted November 21, 2014, http://www.wpafb.af.mil/library/factsheets/factsheet.asp?id=9492.

propulsion, mobility, guidance and control, navigation, energy conversion, materials and structures, and personnel support. Program elements in this category involve pre-Milestone A efforts.[3]

AFOSR accomplishes its mission by investing in basic research efforts for the Air Force in relevant scientific areas. As emphasized by the Air Force Scientific Advisory Board (AFSAB),

> AFOSR is a vital resource to the Air Force and the Nation. It is one of only a few government organizations that funds high-risk, high-payoff research, and the only such organization devoted solely to basic research in support of USAF science and technology needs. Its aggressive approach to cultivate and fund high-risk, potentially high-payoff research is critical for the long-term development of technologies to ensure that the Air Force maintains its technological and competitive advantages over current and future adversaries.[4]

A recent study by the Institute for Defense Analyses Science and Technology Policy Institute (STPI) identified 28 mission-relevant capabilities that have basic science and technology needs that could be addressed by AFOSR in areas such as (1) command, control, intelligence, surveillance, and reconnaissance; (2) component systems and technologies; (3) vehicles (aircraft and spacecraft); (4) human-machine systems; and (5) design through operations.[5] Key functions and attributes of the AFOSR mission include the following:

1. *Broad and deep situational awareness.* Science and technology (S&T) horizon scanning, both domestically and internationally, can ensure that the Air Force is aware of and can capitalize on the latest advances across the full spectrum of S&T fields.
2. *Long-term focus and persistence in funding basic research topics of relevance to the Air Force.* This is especially important now, given that industry is pulling back from longer-term research and development (R&D) investments.
3. *Support of interdisciplinary research.* Scientific breakthroughs increasingly occur at the interfaces among traditional scientific disciplines. Air Force basic research support is managed within a single office, so that AFOSR can effectively facilitate interdisciplinary investigations. Multidisciplinary University Research Initiatives (MURIs), often involving collaboration among the DoD basic research offices, are noteworthy for their successes, according to the Office of the Secretary of Defense (OSD).[6]
4. *Agility in funding.* Some current fields may diminish in importance, and new fields may emerge. AFOSR needs to have the ability to judge quickly whether the Air Force should lead, collaborate, or monitor particular fields.
5. *Effective talent scouting.* Identifying and funding promising researchers and building communities can have lasting impacts on Air Force capabilities. The Air Force can draw from this pool of talented researchers to collaborate with Air Force Research Laboratory (AFRL) and others in the national security community.

Central to AFOSR's strategy is the transfer of fundamental knowledge gained from basic research to the other eight directorates of AFRL responsible for applied research and advanced technology development leading to acquisition programs, to other defense laboratories, to industry, and to the academic community. The benefits of Air Force basic research include not only a more effective Air Force but, more broadly, enhanced national and economic security.

The basic research mission requires continuous and specialized engagement with the academic community to foster intellectual thinking on long-term Air Force challenges. The size and complexity of the global academic base is increasing, and opportunities to harvest scientific knowledge to meet Air Force goals also are increasing.

---

[3] National Research Council, *Assessment of Department of Defense Basic Research,* The National Academies Press, Washington, D.C., 2005.
[4] Air Force Scientific Advisory Board, *Air Force Office of Scientific Research Science and Technology Review*, 2014.
[5] STPI, *AFOSR Portfolio and Gap Analysis,* 2014.
[6] J. Belanichm et al, Institute for Defense Analysis, *DoD's Multidisciplinary University Research Initiative (MURI) Program: Impact and Highlights from 25 Years of Basic Research*, September 2014, https://www.ida.org/~/media/Corporate/Files/Publications/IDA_Documents/STD/D-5361.pdf?.

These and other factors led to this workshop to help ensure that AFOSR is postured to exploit the growing scientific base to meet more complex, long-term needs of the Air Force. AFOSR (and the Air Force, more broadly) need the right people, processes, and tools to effectively harvest knowledge amid the global explosion in research while simultaneously being efficient in light of declining budgets. This workshop focuses on AFOSR's existing management and business practices and sheds light on similar organizations' processes to allow AFOSR to benchmark best practices from the Army, Navy, OSD, and other agencies.

ORGANIZATION OF THE WORKSHOP

In this context, the Deputy Assistant Secretary of the Air Force for Science, Technology, and Engineering requested in 2014 that the Air Force Studies Board of the National Academies of Sciences, Engineering, and Medicine undertake a workshop to address the Air Force's approach to managing its basic research portfolio.[7] The Academies approved the workshop terms of reference and appointed the Committee on Improving the Air Force Scientific Discovery Mission: Leveraging Best Practices in Basic Research Management: A Workshop.[8,9] The committee planned and participated in the workshop and prepared this report summarizing the observations of the numerous workshop participants over two workshop sessions held in April and May 2015.[10] The report summarizes the views expressed by individual workshop participants. While the committee is responsible for the overall quality and accuracy of the report as a record of what transpired at the workshop, the views contained in the report are not necessarily those of all workshop participants, the committee, or the Academies.

Two workshop sessions were held, on April 27-28, 2015, and May 27-28, 2015, in Washington, D.C., during which speakers were asked to address such topics as the following:

1. What are your expectations of AFOSR?
2. What process is used to establish Air Force basic funding requirements?
3. What types of metrics are used to assess AFOSR mission accomplishment?
4. How are manpower requirements established for AFOSR?
5. What are the key elements affecting AFOSR performance?
6. How is basic research transitioned from AFOSR to the Air Force and Department of Defense?
7. How does AFOSR measure the quality, relevance, and impact of its portfolio?
8. Do AFOSR program officers have adequate resources to perform their duties?
9. Is AFOSR funding sufficiently stable; are contracting, obligation, and expenditure policies appropriate?
10. What are AFOSR's principal strengths and opportunities for improvement?
11. What would you change about AFOSR and its relations with other parts of the Air Force?
12. What business practices in your organization should be considered for use by AFOSR?
13. How do AFOSR grantees view AFOSR practices?

MANAGEMENT PRACTICES CONSIDERED IN THE WORKSHOP

While not all-inclusive, the following list provides a flavor of the topics discussed by the workshop participants: (1) organizational culture (e.g., collaborative, risk-taking); (2) senior management (e.g., continuity, background, grade); (3) recruitment, hiring, and retention of well-qualified program officers, including Intergovernmental Personnel Acts (IPAs); (4) effective utilization of program officers (administrative assistants, attendance of pro-

---

[7] Department of Defense (DoD) research, development, test, and evaluation (RDT&E) budget activities are generally categorized as follows: 6.1, basic research; 6.2, applied research; 6.3, advanced technology development; 6.4, advanced component development and prototypes; 6.5, system development and demonstration; 6.6, RDT&E management support; and 6.7, operational system development. See DoD, *Report of the Defense Science Board Task Force on Basic Research*, Washington, D.C., 2012, http://www.acq.osd.mil/dsb/reports/BasicResearch.pdf.

[8] Appendix A provides the workshop terms of reference.

[9] Appendix B provides short biographical sketches of the committee members. The committee was comprised of experts with backgrounds in government basic research, academic basic research, industry basic research, and congressional oversight of basic research.

[10] Appendix C lists the speakers and participants.

fessional conferences, library access); (5) grant selection (criteria, processes, peer review, flexibility, turnover, duration, interdisciplinary); (6) contracting (timeliness, flexibility); (7) financial management (funds availability, obligation and expenditure requirements); (8) timely Institutional Review Board (IRB) determinations; (9) business information (tools, processes, products), workflow automation, and information technology support; (10) technology transition and transfer (processes, building ties, outcomes); (11) national and international collaborations; and (12) measures of success, reputation, and visibility.

RECOMMENDATIONS FROM PREVIOUS STUDIES

The management of Air Force and DoD basic research has been the topic of numerous reports over the years. Excerpted below are a handful of recommendations from several studies completed in the past 10 years that are related to the scope of this workshop. While this workshop report itself does not provide recommendations, the following recommendations from previous studies help to provide additional context for this workshop by illustrating that the topic of managing defense-related basic research is not a new one.

Assessment of Department of Defense Basic Research (2005)[11]

DOD needs to realign the balance of its basic research effort more in favor of unfettered exploration. Senior DOD management should support long-term exploration and discovery and communicate this understanding to its research managers. (p. 2)

DOD's personnel policies should be provide for continuity of research management with managers having an adequate level of authority. DOD should also include within the attributes it assigns to the management of its basic research the discovery of new fundamental knowledge, flexibility to modify goals and approaches, freedom to pursue unexpected paths and high-risk research questions, minimum requirements for detailed reporting, open communications, freedom to publish, unrestricted involvement of students and postdoctoral fellows, no restrictions on nationality of researchers, and stable funding. (p. 2)

S&T for National Security (2009)[12]

Focus on funding people before projects. The "payoff" to DOD is a cadre of people in the internal and external communities who are cognizant of both DOD needs and the forefronts of science, as well as the research itself. (p. 31)

Eliminate large fluctuations in 6.1 funding and schedules. Long-term research efforts cannot be turned on and off with yearly budget cycles and service rotations. Indeed, for a researcher, stable funding is more productive than more variable funding. Pressures to shape the basic research program around the "War of the Month" should be avoided. (p. 32)

Create a basic research advisory committee reporting to the USD(AT&L). The membership of this committee should include the *DDR&E* and appropriate Service personnel, together with an equal number of external members with high scientific and technical credentials from academia and industry. The committee would review and advise annually on the health of DOD basic research. (p. 34)

Report of the Defense Science Board Task Force on Basic Research (2012)[13]

DOD basic research program office directors should rotate active researchers from academia, industry, and FFRDCs using the IPA or HQE programs as appropriate. A useful goal may be to use these tools to keep the average time away from the laboratory low; less than five years for program managers if possible. Tours should be for nominally four years to best match up with the typical rotation of three-year grants. (p. 92)

---

[11] National Research Council, *Assessment of Department of Defense Basic Research,* The National Academies Press, Washington, D.C., 2005.
[12] DoD, *S&T for National Security,* JSR-08-146, May 2009, http://www.dtic.mil/dtic/tr/fulltext/u2/b360036.pdf.
[13] DoD, *Report of the Defense Science Board Task Force on Basic Research,* Washington, D.C., 2012, http://www.acq.osd.mil/dsb/reports/BasicResearch.pdf.

DOD basic research program office directors should provide funds and time for basic research program managers to attend relevant professional society meetings, both in the United States and overseas. These conferences provide excellent opportunities for performer meetings. In addition, program managers should fully participate in professional society activities, including publishing review articles and serving as editorial board members of professional journals. These and other activities enhance the skills and professional reputation of both the program and the program manager, and should be given great weight in the annual evaluation process and in promotion consideration. (p. 93)

DOD basic research program office directors should provide an adequate number of S&T program assistants to help execute the administrative activities associated with proposal review, grant administration, workshop organization, and other program management duties. Assistance with administrative tasks is needed to allow each program manager to perform at their best and to reserve adequate time for higher level activities. Program assistants should have degrees in science, technology, engineering, or mathematics. (p. 93)

AFOSR Science and Technology Review (2014)[14]

AFOSR is a vital resource to the Air Force and the Nation. It is one of only a few government organizations that funds high-risk, high-payoff research, and the only such organization devoted solely to basic research in support of US Air Force science and technology (S&T) needs. Its aggressive approach to cultivate and fund high-risk, potentially high-payoff research is critical for the long-term development of technologies to ensure that the Air Force maintains its technological and competitive advantages over current and future adversaries. (p. 3)

Issue: Funding and staffing are not necessarily driven by a strategic vision aligned with future needs of the USAF. In particular, important research efforts are being reduced due to PO vacancies (e.g., Dynamics & Control; Optimization & Discrete Mathematics; Sensing, Surveillance, and Navigation; and GHz-THz Electronics). *Recommendation*: These programs must be managed to prevent loss of critical capability and allow new proposals to be funded within these important research areas. AFOSR management are encouraged to develop and assess various options, including: 1) manage redistribution of the portfolios that do not have POs among the current POs, and 2) Recruit POs from AFRL TDs, Military, Air Force Institute of Technology (AFIT), US Air Force Academy (USAFA), and Intergovernmental Personnel Act (IPA) program. (p. 6)

Issue: Insufficient PO development and management. The POs are the most important AFOSR resource. They have very specialized skillsets, and they serve a unique role interfacing with the S&T community. However, retirement eligibility, the potential move of AFOSR, and hiring difficulties have the potential to be highly disruptive to the team of POs and their S&T portfolios, and ultimately, the AFOSR mission. *Recommendation*: AFOSR should develop a strategic plan for PO development, management and retention. This includes establishing a more effective process for mentoring and monitoring PO development and rewarding success. AFOSR management should also require POs to periodically refresh the visions for their portfolios based on AF technical strategic plans and the research being done in the broader S&T community. (p. 6)

AFOSR Portfolio and Gap Analysis (2014)[15]

More agreement on AFOSR remaining based in DC and kept independent of other AFRL Directorates (minority of dissenting views). (p. 18)

Some interviewees proposed that AFOSR define and pursue a set of Grand Challenges as a way of both pushing truly breakthrough research and building closer ties between AFOSR and AF operational organizations. (p. 18)

Universal agreement that AFOSR's success (both in portfolio creation and transfer to the rest of the Air Force and DoD) depends on the quality of its program managers (and that current tools for recruitment and retention had serious problems). (p. 19)

---

[14] Air Force Scientific Advisory Board, *AFOSR Science and Technology Review,* 2014.
[15] STPI, *AFOSR Portfolio and Gap Analysis,* 2014.

2

Workshop Participant Observations and Key Themes

AFOSR'S MISSION

Key Theme. Several participants repeatedly stated that AFOSR is a unique organization fulfilling a mission important to national security by connecting the broad relevant basic research community to national security needs in the near and far term.

The evolving U.S. national security environment demands greater scope and depth from science and technology. The Air Force, more than ever before, requires the products of basic research, which are critical to future success. It is essential that fundamental research supported by the Air Force Office of Scientific Research (AFOSR) map to Air Force strategy and that the basic research community be wholly embraced for its long term focus. Figure 2-1 provides an illustration of AFOSR's strategy to manage Air Force basic research

Air Force science and technology must expand at an accelerating rate in order to keep pace with increased complexity of the Air Force mission set and the accelerating spread of relevant technologies to potential adversaries. The emergence of novel science and technology areas will create new threats and opportunities not traditionally considered mission critical by the Air Force. The Air Force continues to perform two fundamental missions: strategic nuclear deterrence and engagement in joint operations. Several decades ago, engagement in joint operations was primarily with kinetic and airlift operations in a single operating domain—creating effects in and from the air. Today, Air Force engagement in joint operations is increasingly demanding, both in scope and in meeting individual mission needs. Joint operations engagement now involves complex operations in and from air, space, and cyberspace with continued and more technically complex kinetic operations and greatly increased intelligence, surveillance, and reconnaissance (ISR); mobility operations; delivery of other space-based capabilities; and non-kinetic operations. Basic research provides the essential underpinning for delivering the needed science and technology to meet these continually expanding demands for Air Force missions. It also provides an essential window onto emerging frontiers in science and technology, both in the United States and internationally, in areas directly relevant to current and future Air Force priorities.

AFOSR has the unique mission of providing the base for science, technology, and engineering growth and development. It connects broadly relevant basic research to national security needs in the near and far term. It provides thought leadership for academia and industry to establish new disciplines that deliver uniquely critical technologies. It supports growth in depth and breadth of the best and brightest researchers, helping them become

FIGURE 2-1 AFOSR technical strategy. SOURCE: Thomas Christian, AFOSR Director, "AFOSR Snapshot," presentation to the committee on April 27, 2015.

leaders in their fields. It establishes strong ties between those future technology leaders and Air Force mission needs. It mitigates the risk of technology surprise and imbues the Air Force with an integrated awareness of both novel offensive measures and orthogonal countermeasures to Air Force capabilities that will materialize from emerging fields of science and engineering.

In the opinion of several workshop speakers, AFOSR international offices are especially critical for building ties with leading researchers throughout the world, obtaining current knowledge about foreign scientific advances, and building international good will. Overseas grants made by AFOSR, although modest in size and number, are particularly welcome in many countries because they usually involve less cumbersome reporting requirements and serve as strong endorsements, often attracting other funding. The AFOSR international program officers are the Air Force Research Laboratory's (AFRL's) "boots on the ground." As noted by one speaker, however, all this is of little value if the information obtained is not shared with the rest of AFRL. AFOSR international offices need the resources to do so.

Several speakers and committee members expressed the view that the collocation of AFOSR with the Office of Naval Research (ONR), the Defense Advanced Research Projects Agency (DARPA), the National Science Foundation (NSF), the Department of Energy's Office of Science, the Office of the Assistant Secretary of Defense for Research and Engineering (ASD(R&E)) basic research office, and other federal funding agencies in the Washington metropolitan area is highly beneficial. The last Base Realignment and Closure Commission made a similar statement,[1] as have many universities and professional societies. Some went on to say that collocating AFOSR with other AFRL directorates would push it toward near-term, risk-averse research, perhaps even putting its continued existence at risk.

On the other hand, most speakers felt that AFOSR support of quality basic research at other AFRL directorates has merit and should be continued. They noted that such research strengthens the qualifications of AFRL researchers, builds ties to the academic community, and facilitates basic research technology transition, the subject of a later theme at the workshop. AFOSR contributes not only funds but also establishes academic contacts and provides quality control. However, block-funding basic research for AFRL technical directorates would not be wise, according to speakers who addressed the issue.

Merging DoD basic research offices has been proposed from time to time, and this topic came up during

---

[1] Defense Base Closure and Realignment Commission, *Final Report to the President*, Vol. 1, pp. 281-282, 2005.

the workshop. Generally, there seemed to be little enthusiasm for it among most of the participants. However, in the opinion of several workshop participants, the JASON committee recommendation to "create a basic research advisory committee reporting to the USD(AT&L)" has merit.[2] "The membership of this committee should include the DDR&E (now, ASD(R&E)) and appropriate Service personnel, together with an equal number of external members with high scientific and technical credentials from academia and industry. The committee would review and advise annually on the health of DoD basic research."[3]

Historically, some participants asserted, AFOSR has performed its responsibilities well, but faces obstacles to continued mission efficiency and effectiveness in meeting future needs. The key themes discussed below provide specific examples of why more attention is needed on facilitating the mission, including program officer numbers and diversity, rapid management turnover, and reductions in mission support.

AFOSR'S MISSION LEADERSHIP

Key Theme. In the opinion of numerous participants, stable, qualified, committed leadership at the top in AFOSR and its mission and support organizations also is essential for effective mission performance. These participants noted that all key functions should report to the AFOSR director.

The need for stable, qualified, committed leadership at the top in AFOSR and its mission and support organizations was strongly emphasized by multiple workshop participants as being essential for effective mission performance. Reductions in AFOSR senior positions have reduced the senior civilian leadership to a single Senior Executive Service (SES) member and a single senior scientific or professional (ST) member, when less than a decade ago the number of SES members was as many as five. This challenges the ability of AFOSR to engage both within the Air Force and with the external technical community at the senior leadership level. Numerous participants asserted that additional SES, ST, or equivalent positions are needed to support enhanced outreach to the basic and applied research and advanced technology development communities.[4] This added investment in leadership could more effectively enhance communications with federal and university science and technology leaders, promote technology transition, and articulate the criticality of basic research to Air Force mission success. Based on comments by several speakers, it appears that AFOSR has a disproportionately small number of SESs for its size as a basic research organization. In the words of one speaker, AFOSR "needs big dogs."

One participant suggested that the AFOSR director and other senior leaders should be expected to occupy their positions for at least 4 years, as has been the case until recently. In contrast, AFOSR has had four directors or acting directors in the past 2 years. As recommended in an earlier review of DoD basic research, "Personnel policies should provide for the needed continuity of research management in order to ensure a cadre of experienced managers capable of exercising the level of authority needed to effectively direct research resources."[5] Further, many speakers believe that AFOSR leadership should, in general, have technical qualifications at least comparable to those of program officers. Perhaps, as suggested by one speaker, national searches should be conducted for future directors.

AFOSR leadership does not have authority over some functions essential to effective and efficient mission accomplishment. A critically important example is the contracting office. The performance of this office is essential to the efficient execution of grants and contracts in alignment with basic research goals and academic timelines (namely, semester and graduation dates). According to former AFOSR directors, it is critical that program officers and contracting personnel work as a team, but separate reporting chains inhibit close, timely collaboration. Others spoke of the "tyranny of the functionals." Consolidation of mission support functions removed from AFOSR has been characterized by many as being in pursuit of efficiency. As noted by one participant, there may be efficiencies in support functions, but when it is at the expense of the mission performance, the priorities are questionable at best. Many participants believe that mission effectiveness should have priority.

---

[2] JASON Committee, *S&T for National Security*, JSR-08-146, Mitre Corporation, McLean, Va, 2009.
[3] Ibid.
[4] Equivalent positions might include DR-5 positions, like those in the Army, Sec 1101 positions, like those at DARPA, or IPAs experienced in science and technology management.
[5] National Research Council, *Assessment of Department of Defense Basic Research*, The National Academies Press, Washington, D.C., 2005.

Several speakers and committee members suggested that financial policies appropriate to the Air Force as a whole may not be appropriate for basic research management. For instance, several speakers stated that failure to obligate most funds for universities in the spring of each year compromises the quality of the funded research. Participants added that the timely obligation of funds could be accomplished by (1) assuring that AFOSR receives all its funds very early in the fiscal year and providing it with sufficient administrative personnel to deploy those grants quickly or (2) allowing AFOSR to carry a significant fraction of its funds into the second year of appropriation. In that light, the requirement that research development test and evaluation (RDT&E) funds be 88 percent obligated at the end of the first year is a "self-inflicted wound" that leads to less-than-optimal deployment of grants. The participants were told that AFRL is seeking an exemption to this and other financial policies for AFOSR. Year-to-year variations in the basic research budget also are harmful, according to several speakers.

AFOSR'S MISSION FORCE

Key Theme. The mission execution force for the work of AFOSR is its cadre of program officers, according to several participants. Therefore, they argued, recruiting and retaining the right number of well-qualified program officers, including through the Intergovernmental Personnel Act (IPA), is extremely important.

Most of the workshop participants supported the idea that AFOSR's effectiveness and efficiency in fulfilling its mission rests on the effectiveness and efficiency of its mission execution force—its cadre of program officers. Several pointed out that the highest-quality program officers—with high levels of technical competence and the leadership and management experience to lead basic research—are essential to the success of AFOSR. These individuals must be recognized thought leaders and technical champions in their fields. They should have a broad and critical understanding of their technical fields in order to provide the essential foundation to frame the Air Force research agenda and attract ideas from the best and brightest, both domestically and internationally. (These same views have been expressed by the Defense Science Board, the Air Force Scientific Advisory Board, the JASON Committee, the Science and Technology Policy Institute, and others.) To paraphrase one of the Air Force chief scientists at the first workshop session, program officers need to be able to speak to Nobel Prize winners one day and a four-star general the next.

AFOSR program officers at the first workshop session tended to agree with the idea that it was beneficial to develop and nurture deep and constant engagement with the technical communities of enduring significance to the Air Force to better understand leading-edge technical concepts. To ensure continued attention to emerging technical opportunities and needs, several participants suggested that the cadre also needs those who rotate into the role for specific periods. To fill this requirement, the cadre needs to be drawn from multiple sources: federally funded research and development centers (FFRDCs) at universities, University Affiliated Research Centers (UARCs), government laboratories, and industry laboratories. Many speakers and committee members emphasized the importance of utilizing the IPA to bolster the AFOSR program officer cadre and facilitate the infusion of new ideas and concepts into the office. Several speakers expressed the view that AFOSR consider increasing turnover in its program officers, perhaps by blending its current model with that of DARPA or NSF to ensure competitiveness, creativity, and transition focus. Some added that AFOSR also might consider exchanging program officers with other basic research funding agencies in order to further build cooperation and absorb best practices from others.

As observed by a senior member of AFRL management at the second workshop session, the diminishing number of program officers, the expanding range of disciplines to be covered, and the increasing rate of innovation combine to limit the effectiveness of AFOSR. Currently, AFOSR is understaffed in program officers from its optimum level by about 25 percent, and important research efforts are being reduced due to vacancies. A number of workshop participants asserted that correcting this situation must be the foremost priority of AFOSR and will require special attention to recruiting and retention.

Some speakers suggested allocating time on a weekly basis to allow program officers to maintain their technical stature by performing, presenting, and publishing research. The Army Research Office and NSF, for example, encourage program officers to undertake these activities. It was noted that AFOSR policy also permits personal research, but little time is available in practice to do so. Most participants hold the view that the most important focus should be to leverage the knowledge, experience, and commitment of program officers by ensuring that the entire enterprise recognizes and treats the program officer cadre as the mission execution force. The function of all other activities must be to make this mission force as successful as possible. Figures 2-2 and 2-3 provide suggestions from two current Air Force major command chief scientists, presented at the first workshop session, related to improving the professional experiences of AFOSR program officers.

- Expand the AFOSR rotational assignment program to offer tailored experiences at operational locations to "blue" AFOSR PM's

- Bringing transition partner presentations into AFOSR workshops to establish pathways and expedite delivery of integrated capabilities by AFRL

- AFOSR can play an important role in integrating research performed by multiple AFRL directorates—especially if focused on a Grand Challenge or enabling integrated capabilities

FIGURE 2-2 Developing people. SOURCE: Janet Fender, Chief Scientist, Air Combat Command, "Improving the Air Force Scientific Discovery Mission: Leveraging Best Practices in Basic Research Management (Personal Views)," presentation to the committee on April 28, 2015.

- **What AFOSR Needs**
  - #1: talented, diverse, entrepreneurial, and empowered Program Managers
  - Visionary, eminent, and respected leaders
  - Agile, multi-capable support offices
  - Consistent and sufficient funding
  - Strong linkages and connections
- **What MAJCOMs Need**
  - Engage and dialog on new technologies and promising breakthroughs—early and often
  - No disruptions to S&T pipeline to fielded systems

FIGURE 2-3 Observations and perspectives. SOURCE: Donald Erbschloe, Chief Scientist, Air Mobility Command, "Perspective on Improving the Air Force Scientific Discovery Mission: Leveraging Best Practices in Basic Research Management," presentation to the committee on April 28, 2015.

AFOSR'S MISSION SUPPORT

Key Theme. Many participants raised the need for practices and processes that facilitate and support discovering and leveraging the best S&T work, wherever it is taking place, and others stated that establishing relationships with the best and brightest in national security work are essential to program officer effectiveness.

Several participants asserted that AFOSR program officers need to focus on the mission. In this context, they noted that every hour spent on paperwork in the office is less time interfacing with the research community, attending conferences, reading journals, and fostering interactions across all the various academic and Air Force stakeholders, customers, and performers. Hence, having adequate administrative support is essential. In this connection, as noted by many, AFOSR's improved electronic work flow system and other internal information technology business processes are certainly valuable in reducing the administrative workload and were viewed as a best practice. But both current and past program officers and AFOSR leadership expressed the view that by eliminating assistant program officers and some other support personnel, administrative efficiency has been allowed to take precedence over mission effectiveness. Nearly every speaker stated that this situation needs to be corrected immediately. This thought has been captured by others, including the Defense Science Board.

> DoD basic research program office directors should provide an adequate number of S&T program assistants to help execute the administrative activities associated with proposal review, grant administration, workshop organization, and other program management duties. Assistance with administrative tasks is needed to allow each program manager to perform at their best and to reserve adequate time for higher level activities. Program assistants should have degrees in science, technology, engineering, or mathematics.[6]

Many participants argued that organization, practices, and processes that facilitate and support discovering and leveraging the best S&T work, wherever it is taking place, are essential to the success of AFOSR. Several added that they are essential to establishing relationships that produce relevant proposals, funded grants, and other interactions with outstanding researchers in the national security arena. There was broad understanding among many of the speakers and participants that, to be effective in performing these functions, program officers must participate in a wide range of conferences, workshops, and visits to centers of excellence. Relationships matter. A recurring concern heard from several presenters was that onerous bureaucratic processes have been institutionalized, restricting the attendance of scientific and technical conferences.[7] Many spoke about the detrimental impacts that current policies and practices are having, not only at AFOSR but across the DoD science and engineering (S&E) community. Further, some participants added, the globalization of technology requires that this participation include interfacing in domestic and foreign venues. Many participants agreed that approval authority for conference attendance could be delegated to the AFRL commander and, whenever possible, to the AFOSR director.[8]

There was also agreement by many participants that program officers must have the broadest practical access to information sources to understand the pace of technology, again, both foreign and domestic. Several program officers stated that they did not have access to the journals needed to perform their work. These program officers also emphasized the growing importance of data analytics in identifying important research opportunities, as did several speakers and committee members. For instance, a committee member with relevant experience stated that "DoD S&T is behind the private sector in making use of data and data analytics. We should be supporting access to data sets and tools that can help them do their jobs more effectively." He added that a technical librarian and others with expertise in big data also are needed. A few speakers and committee members expressed concern over the lengthy delays involved in obtaining Air Force approval of human and animal use research protocols.

---

[6] Defense Science Board, *Report of the Defense Science Board Task Force on Basic Research,* 2012.
[7] Government Accountability Office, *Further DOD and DOE Actions Needed to Provide Timely Conference Decisions and Analyze Risks from Changes in Participation*, GAO-15-278, Washington, D.C., March 2015, highlights page.
[8] Report to Accompany S. 1376, "National Defense Authorization Act for Fiscal Year 2016," Report 114-69.

TRANSITIONING MISSION SUCCESS

Key Theme. Multiple participants at both workshop sessions cited that technology transition from AFOSR relies excessively on personal relationships established by program officers, and several claimed that a more formal and focused approach to transition is needed.

Many speakers and committee members stated that AFOSR' focus on basic research, with other directorates within AFRL focused on applied research and development, is a desirable and successful model to sustain robust basic research. But several committee members and workshop presenters expressed concern that this model can leave the process for transition undefined, ad hoc, and excessively dependent on personal relationships and the initiative of the individual program officer. (In contrast, other speakers felt that personal relationships were an effective and desirable mechanism for fostering transition.) The expectation is that many AFOSR-funded basic research results should lead to future operational mission capabilities. There are certainly numerous examples to validate this expectation. For this to occur, in the view of many participants, there needs to be an orderly process for movement to applied research and eventually to industry activity to exploit the science and technology successes. A sentiment expressed by many participants is that fostering relationships between AFRL leadership and the combatant command and major command (MAJCOM) chief scientists is essential. Figures 2-4 and 2-5 provide suggestions from two MAJCOM chief scientists on potential game changing research areas for AFOSR investment that were presented at the first workshop session.

Several speakers suggested that AFOSR consider approaches to strengthen critical connections in novel ways, such as establishing a transition office, holding focused workshops, and exchanging personnel with industry and other parts of DoD. Of course, they added, such initiatives would require additional personnel. In fact, one program officer remarked that sufficient time no longer was available for effective interactions with counterparts in other parts of AFRL.

Interestingly, a few AFOSR program officers described DARPA as more receptive than AFRL to transitioning AFOSR discoveries. Several participants believe that AFOSR needs to "market" its successes better. One such approach suggested is showing a timeline of the transition of basic research to applied research and advanced technology development and capability delivery. In this context, one participant stated, a history of game-changing investments initially conceived and funded by AFOSR would be particularly valuable in communicating the impor-

- **Recommendation: energize innovation by focusing multi-disciplinary research on Grand Challenges for the AF**

- **Proposed Grand Challenge: enable low SWaP, ultra precision, stable timing**
    - Address a major gap for operations in contested environments
    - A landmark accomplishment for AFOSR
    - Revolutionize AF operations

- **Engage AFRL Directorates, NIST, DARPA, NRL, academia and industry to develop game changing capabilities**

FIGURE 2-4 AFOSR grand challenges. SOURCE: Janet Fender, Chief Scientist, Air Combat Command, "Improving the Air Force Scientific Discovery Mission: Leveraging Best Practices in Basic Research Management (Personal Views)," presentation to the committee on April 28, 2015.

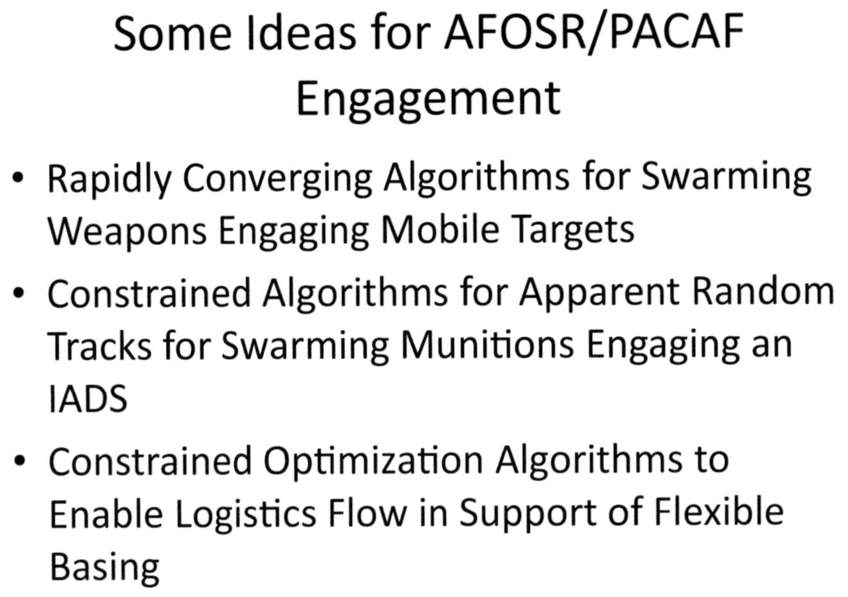

Some Ideas for AFOSR/PACAF Engagement

- Rapidly Converging Algorithms for Swarming Weapons Engaging Mobile Targets
- Constrained Algorithms for Apparent Random Tracks for Swarming Munitions Engaging an IADS
- Constrained Optimization Algorithms to Enable Logistics Flow in Support of Flexible Basing

FIGURE 2-5 AFOSR grand challenges. SOURCE: Azar Ali, chief scientist, Pacific Air Forces, "Improving the Air Force Scientific Discovery Mission: Leveraging Best Practices in Basic Research Management (Personal Views)," presentation to the committee on April 28, 2015.

tance of basic research and fostering transition to higher-level research and development. One example cited by the executive director of ONR at the second workshop session is how ONR uses its Naval Reserve Unit to document successful transitions and mine lessons learned to improve the transition process. There was not an expectation by the workshop participants that all basic research will transition, but given the speed of technology development today, several noted that it is important to develop a climate where new ideas generated can be brought to fruition more rapidly. A speaker suggested that AFOSR sponsor a few basic research grand challenges. The Small Business Technology Transfer (STTR) program often served in the past as an important vehicle for transitioning discoveries to industry for further development, according to several speakers. Whether this remains the case, now that AFOSR no longer manages this program, is unclear.

PERSPECTIVES FROM A FORMER U.S. AIR FORCE CHIEF OF STAFF

Larry Welch was the 12th chief of staff of the Air Force. Box 2-1 provides his personal thoughts on AFOSR as a national security organization, which reflect on the presentations and discussions over the two workshop sessions.

> **BOX 2-1**
> **AFOSR as a National Security Organization**
>
> *Larry D. Welch, Senior Fellow, Institute for Defense Analyses*
>
> The workshop heard a widely held view from presenters and participants that AFOSR is a unique organization fulfilling a mission important to national security. Holders of this view included the Air Force operational major commands, Office of the Secretary of Defense organizations involved in basic research, representatives from the Air Force acquisition community, and the Army and Navy basic research organizations. AFOSR serves the role of connecting the broad relevant basic research community to national security needs in the near and far terms. To perform this mission, AFOSR must have broad and deep knowledge of what is going on nationally and globally in basic research, where it is being done, and who is doing it. They must also have broad and deep awareness of Air Force needs, including long-term strategic plans, priority gaps in current capability, ongoing programs at the 6.2 (applied research) and above level to fill those needs, and issues with fielded systems that could require 6.1 work. Using this knowledge of basic research work and sources and the broad and deep set of Air Force needs, AFOSR connects the relevant source to the Air Force need. I am aware of no other organization that serves this purpose or has the requisite focus and experience set to serve this purpose for the Air Force at the basic research level.
>
> The mission execution force for the work of AFSOR is the cadre of program officers. The efficiency and effectiveness of AFSOR in fulfilling its mission rests on the effectiveness and efficiency of the mission execution force—the program officers. Strongly held views from AFOSR program officers, past AFSOR directors, and others who attended the workshop were that important best practices have been compromised by cost-cutting masquerading as efficiencies. These departures from best practices are the antitheses of efficiencies. Instead, they transfer inappropriate workload to program officers detracting from the performance of their mission of understanding priority needs, finding the best sources to address those needs at the basic research level, funding the work, managing the portfolio of that work, and facilitating the transition where appropriate to research and development beyond 6.1.

3

Synopsis of Workshop Sessions

Listed below, in chronological order, are short abstracts or summaries of remarks provided by workshop speakers. The actual presentations were, of course, much more extensive and often covered important issues not described in the abstracts.

SESSION 1: APRIL 27-28, 2015

Deputy Assistant Secretary of the Air Force for Science, Technology, and Engineering—David Walker (SES)

David Walker (SES) began by stating that travel restrictions have been relaxed somewhat and that he feels restrictions are partly an "urban legend; he wants to delegate conference approval authority to technology directorates (TDs). AFOSR is Dr. Walker's primary outreach to the international community and he does not want to hamper that by travel restrictions; this is a critical mission. Dr. Walker stated that AFOSR has gone through a lot of change over time. Over time the Air Force has tried to integrate AFOSR into the overall AFRL portfolio and strategy (i.e., 12 mission focuses). The Air Force wants to maintain a reasonable funding level for AFOSR; the target is to maintain funding of 61102F at no less than 15 percent of AFRL budget. From an AQR perspective, AFOSR has to provide long-reaching, wide aperture research and identify opportunities we do not know about. Further, AFOSR has the ability to go out and build connections between AFRL and the basic research community. According to Dr. Walker, 30 percent of 61102F funding is spent on in-house Air Force research and that this research is well-linked to the extramural program (e.g., centers of excellence). Dr. Walker stated that the FY15 budget for Air Force basic research was increased by $75 million by Congress. Dr. Walker then asked the question, how do we get ideas from AFOSR and academia into operations? Further, he wants to link research to Air Force core functions and strategic plans in the same way that there is a tight relationship between advanced technology development research and operational gaps. Dr. Walker expects AFOSR to be cognizant of Air Force problems and to align its portfolio accordingly—for example, key areas such as assured communications; positioning, navigation, and timing (PNT), cyber, high-temperature materials, and directed energy (DE).

Air Force Research Laboratory—Maj Gen Thomas Masiello, Commander

AFOSR is an integral part of the Air Force's single Science and Technology (S&T) Enterprise, the Air Force Research Laboratory (AFRL). AFRL guides and focuses all S&T investment for the Air Force including basic and applied research and advance development. We conduct in-house research, partner with other government labs, and stimulate S&T in industry and academia to ensure warfighters get the best possible S&T to respond to urgent operational challenges, address the highest priority Air Force needs, and revolutionize future Air Force operations. AFOSR supports these efforts by executing a basic research investment plan that compliments the identified basic research gaps in AFRL Technology efforts and invests in basic research areas key to enabling the Air Force Vision as defined by the Secretary of the Air Force (SECAF) and the Chief of Staff of the Air Force (CSAF).

AFRL Expectations of AFOSR

The AFOSR mission is to Discover, shape, and champion basic science that profoundly impacts the future Air Force. This means they are to protect the Air Force from technological surprise and to ensure our basic research technology base remains at the cutting edge of the possible. To do this AFOSR is expected to understand the Air Force vision, evolving long term challenges, and priorities. Identify best opportunities for significant scientific advancements and breakthrough research around the world. Rapidly bring the right researchers (from around the world) and resources (Air Force and partners) to foster revolutionary and transformational basic research for Air Force needs to enable the AF to exploit these opportunities through transition of revolutionary S&T (to AFRL TDs, other DoD entities, and industry) that support the AF supply chain.

Alignment of the AFOSR Strategic Plan with AFRL Goals

The AFOSR Strategic Plan clearly articulates how AFOSR will accomplish its part of the AFRL mission in a manner that is grounded in the success criteria listed above. Specifically it reflects a strong understanding of the Air Force Vision and it long term challenges. It identifies the best opportunities for significant basic research advancement for our airmen. The AFOSR strategy outlines how it will access and partner with the world's best subject matter experts so that we can leverage their knowledge and resources to fulfill Air Force needs. Finally it promotes an awareness of opportunities to transition basic research to the AFRL Technical Directorates, industry, or other government agencies.

Mechanisms Used by AFRL to Stay Abreast of Emerging Scientific Discoveries Funded by AFOSR

First AFOSR's basic research contribution to the environmental scan accomplished yearly as part of the AFRL PPBE process and the POM initiation process provides insights of gaps and overlaps with AF funded research. AFRL Technology Directorates are represented on the teams which assess the status of all AFOSR research during the Annual Spring Review. The results of the Spring Reviews are made available to all scientist and engineers throughout AFRL. Significant accomplishments resulting from AFOSR funded basic research is provided to the entire AFRL workforce in the monthly activity report (MAR). All AFOSR developed technical reports are made available for use by all AFRL employees through DTIC. In addition to these more formal methods of communicating the results and status of AFOSR sponsored research there is continual communication between AFOSR program managers and researchers in the AFRL Technology Directorates.

Metrics Used to Access AFOSR Mission Accomplishment

The SECAF chartered Air Force Scientific Advisory Board (SAB) reviews AFOSR from a quality and relevancy perspective at least every other year. Quality is assessed by expressed as a percentage of the AFOSR sponsored efforts which meet or exceed the expectations of the SAB members. Relevancy is assessed as a percentage of the AFOSR sponsored efforts which in the view of the SAB have a solid tie to the Air Force vision, its long term challenges, or new opportunities the Air Force could exploit to its advantage. The SAB also assesses the percentage of AFOSR efforts it feels are appropriately funded.

Internal to AFRL the Research Council provides feedback on the alignment of AFOSR funded efforts with those of the Technology Directorates. Technology Directorate participation in the AFOSR Spring Reviews provides direct feedback from the internal AFRL researchers to the AFOSR program managers. The financial health of AFOSR is assessed using the same metrics used to assess the Technology Directorate in compliance with directives from higher headquarters.

Role Played by AFRL in Establishing AFOSR Financial, Contracting, and Personnel Policies

As an integral member of AFRL, AFOSR's financial, contracting and personnel policies are aligned and controlled by HQ AFRL. AFRL/PK, AFRL/FM and AFRL/DP are the respective Center Senior Functionals (CSF) for all personnel, financial management, and contracting personnel across the Air Force Research Laboratory, including AFOSR. As CSFs they ensure that all personnel, financial management, and contracting positions are properly trained. FM provides guidance related to the implementation of financial tools which ensure financial data integrity across the labs. PK provides guidance concerning the establishment and execution of assistance and contractual instruments across the Center. DP provides guidance related to all personnel matters including hiring, firing, promotions, appraisals, and employee relations.

In their role as the CSF they disseminate and provide additional guidance for DoD, Air Force, and AFMC policies, as well as providing AFRL specific policies. To improve alignment with AFRL policies financial servicing was moved to the 88th ABW at WPAFB to ensure proper accounting controls beginning in FY14. Similarly to comply with Air Force Personnel Center realignments the Customer Service Representative (CSR) servicing of AFOSR personnel has been transferred to the 88ABW at WPAFB.

How Does Research Transition from Single-Investigator-Sponsored Grants into AFRL-Supported Efforts? How Does This Compare to Basic Research Performed by In-House Researchers?

While there is no formal process for the transition of single investigator grant research and in-house research there are many transition paths for single investigator sponsored grants to an AFRL sponsored lab effort. The key to transition is ensuring the research results are socialized through reports and one on one interactions between the PI and the TD S&Es leading applied research/advanced technology development research efforts. Investigators can transition/sell Intellectual Property to entities already in the AF supply chain, Investigators can become suppliers to entities that already are in AF supply chain, or AFRL personnel can seek out single investigators with demonstrated basic research to address their S&T Gaps. The likelihood of successful transition of in-house basic research is greater for extramural work, because the path is more defined as it is normally tailored to address specific applied research/advanced technology development plans. Multiple transition paths enable smart balancing between basic research Tech Push and basic research Tech Pull.

Air Force Office of Scientific Research—Thomas Christian (SES), Director

Dr. Thomas Christian stated that AFOSR has funded 78 Nobel laureates over time. There is a host of strategic guidance that AFOSR follows. AFOSR believes in the "find, fund, forward" process (see Figure 1-1). Christian noted that AFOSR spends approximately 90 percent of its core funding domestically, but has offices overseas to better locate pertinent international research (London, Tokyo, Santiago). Efforts have been made to consciously integrate domestic and international efforts. According to Christian, AFOSR has only 32 program managers and wants to reach 40. In his mind, international programs officers are "forward deployed" and are "boots on the ground." With respect to personnel and leadership, Dr. Christian noted that there have been four directors/acting directors over the past 2 years

AFOSR Basic Research Division—Col Robert Kraus, Chief

Col Robert Kraus noted that AFOSR business practices are now more digital, utilizing the AFRL enterprise business systems, whereas there used to be much more paper involved. According to Kraus, there are nine steps in the grant award process. AFRL uses four business systems to promote transparency and traceability. It takes 90 days to award a grant once a program officer recommends a proposal for funding. Required expenditure rates are burdensome for AFOSR, and it is seeking relief (with AFRL's blessings). Although the enterprise business systems streamlined some processes, there has been an increase in administrative work for program officers (POs), which detracts from focusing on research. AFOSR, and the Air Force as a whole, has lost funding for administrative support; AFOSR program officers have had to shoulder additional administrative responsibilities, such as arranging conference approvals and coordinating proposal reviews. Additionally, AFOSR needs a technical librarian to do literature searches, etc. Kraus noted that AFOSR is doing a lot of work to determine what analytics and metrics are useful. It is hoped that this will better inform spending priorities. Specifically, he noted, AFOSR has a science analytics working group that will evaluate those metrics that work well and that drive behaviors. Finally, Kraus noted that AFOSR is making use of social media, such as Facebook and YouTube.

Finding Panel

The panel for the discussion with AFOSR program staff on "finding" included Dr. John Luginsland, program officer, laser and optical physics; Dr. Enrique Parra, program officer, ultrashort pulse laser matter interactions;

Lt Col Victor Putz, program officer, European Office of Aerospace Research and Development (EOARD) physics; and Maj Justin Silverman, staff judge advocate. The panel addressed the topic of "finding" Air Force-relevant research—for example, access to electronic library resources, conference participation, sponsoring workshops, participation in professional societies, social media, and analytics, among others.

Lt Col Victor Putz, Program Officer, EOARD Physics, AFOSR/IOE

Coming in at a time of relative austerity, I made the point that analytics and bibliometrics were excellent tools for a high-level "survey" view of the research landscape, but that they were in no way a substitute for strong subject-matter experts. Since 'organic' scientists, mathematicians, and engineers (SMEs) as POs were valuable, relationships with other SMEs (such as current or past PIs) are valuable for both awareness of current research, research funding opportunities ("what's your great but risky idea that no one else wants to fund?") and pointers to other researchers ("who else should we be talking to?") Experts being shown as valuable resources, getting them together in one place (attending and funding conferences) is an extremely cost- and effort-efficient way to get very current and "live" views into the state of research. Gen Welch asked Dr. Parra why one would need both journal access AND conference attendance. Dr. Parra's point was that of time scales; journal access and bibliometrics will give you access to what *has* been going on (and a "larger-scale" view of the field), and conferences and experts will give you shorter-term "current" data on very specific directions of research. The analogy I wish I'd come up with for that is the idea of going on a road trip: Would you rather have a map (journals/analytics, showing you the whole path but fuzzy on details) or be able to stop and ask for directions (conferences/experts, giving you fine/close details but no large view)? Of course you would want both. I did not say this during the panel, though. Journal access remains a problem, although we are very glad to have the access we do. Time for journals and reading seems to be a slightly growing issue, with POs now doing more contract administration and less "looking for science."

I made the comment that we needed items both of relevance to the Air Force and that appear to be good science, but that still represented a "risk spectrum"—only choosing grants that directly aligned with current AFRL research projects would be "safe and useful," but would likely result in evolutionary improvements rather than "risky" investments that might potentially address a mission function and could produce nothing at all . . . or a revolutionary capability. From an international perspective, some mechanical parts of the process dictated by higher offices (grants.gov, the System for Award Management (SAM) registration) cause difficulties with international registrants that result in a great deal of frustration. Similarly, while we *can* fund workshops and conferences, we have enough restrictions (in how money can be spent, etc.) that applicants often get very frustrated. International offices still face some difficulties with foreign universities being distrustful of accepting U.S. military funds (the anecdote was an application from the University of Dresden, which wanted some sort of legal assurance that nothing that resulted from the grant could ever be used in weapons—obviously an unfulfillable requirement).

John W. Luginsland, Program Officer, Laser and Optical Physics, AFOSR/RTB

I was on the "Find" mission panel. Much of the finding panel discussion centered about how/why we thought we were finding the best people and projects to fund. The answers to this split into two main discussion threads: conferences and analytics. First, the critical need to go to conferences, which is the traditional path for finding high quality research, was emphasized. Conferences provide a number of advantages, including: 1. finding out about breakthroughs months to years before a result is published in the archival literature, 2. getting to see the community's response to your PIs and their research, such as seeing who collaborates with whom, and 3. being introduced to new ideas you didn't even know you were interested in. Second, the discussion highlighted the important potential of analytics to help POs sort through the deluge of data from the published literature, at the least, and with technical progress in the science of analytics, potentially identify new researchers via analytics in the future. Dr. Parra used the term "cognitive bias" as something to be fought in finding new people, and I think that both conferences and analytics support this fight against cognitive bias. There was discussion of how easy/hard it is to go to conferences, and some discussion of how we interact with researchers in the other TDs - conferences play a role here, too, but there was also an important discussion of AFRL/AFOSR PO introducing their PIs to AFRL TD staff, and vice versa. There was some discussion of how individual POs work more or less closely with AFRL's other TDs, and how these personal interactions are important, but there is a lack of formal procedures to drive these interactions. This lack of formal produces does have a benefit in offering flexibility as the wide variety of fields that AFOSR funds have different cultures and collaboration styles. Finally, I would have liked to have a more robust discussion of AFOSR's interaction with professional societies, which I believe can play a critical role in the find mission. However, I think we have some internal work on how to best interact and serve with the professional societies while maintaining and meeting the ethics standards from JA. While there is a strong internal component to this, it might be good for the committee to weigh in on if they believe the interaction and professional development that the societies offer is important, and in what ways.

Funding Panel

The panel for the discussion with AFOSR program staff on "funding" included Mr. Mark Amundson, program element monitor for basic research (AQRT); Dr. Michael Berman, program officer, molecular dynamics and theoretical chemistry; Mr. Phillip Cherbaka, director, information technology and chief information officer; Mr. Rickey Lawrence, chief of finance; and Ms. Dorothy Howe, deputy chief of contracting. The panel addressed the topic of "funding" Air Force-relevant research—for example, impact of sequester on planning process, budget uncertainty, budget execution, co-funding mechanisms with AFRL, other Services or agencies, requirements driven by contracting policies, OSD, and Air Force expenditure or obligation goals.

Michael R. Berman, Ph. D., Program Officer, Molecular Dynamics and Theoretical Chemistry, AFOSR/RTE

During the session with the Air Force Studies Board last week, Gen. Welch asked the question, "How do we know if we are attracting the best people to propose to AFOSR?" I wanted to provide the following input. Many AFOSR programs have developed positive reputations in the scientific community, so that in addition to POs going out to find the best, top scientists come to us as well, based on our opportunities and our program's reputations. For example, this year my program had inquiries from 25 excellent scientists about our Young Investigator Program. The following excerpt from an email that I just received demonstrates this point, and I think it is instructive: "I am just starting as an assistant professor at Columbia University and I wanted to touch base with you since both Al Wagner (DOE) and Elaine Oran (Naval Research Laboratory [NRL]) had independently suggested I start a dialogue about your program. In case you would find it helpful, I have included a brief description of my background and my research interests below this email; my CV attached and website below also provide additional information."

Rickey Lawrence, Chief, Finance, AFOSR/RPF

The issue surrounding the poor expenditure rates in relation to basic research stem from several issues. The first is the move away from scheduled payments in FY2011 to invoicing, which created a problem because it took time for many universities to get registered in Wide Area Workflow (WAWF), an Air Force invoicing system. Second, grants are not subject to the Prompt Payment Act, meaning that if an invoice isn't paid within 30 days, the Air Force incurs interest. Third, the huge backlog of cash-on-hand at universities, combined with their failure to put these funds in interest-bearing accounts (which drove the Air Force to change to invoicing), had resulted in millions of dollars sitting on grants that had to be paid through WAWF before later year funds could be disbursed. For example, an FY2013 grant that has two option years may still have unexpended FY2013 funds on it, even though additional FY2014 and FY2015 funding may have also been obligated. The university is required to finish expending FY2013 funding before it can start expending FY2014 and FY2015 option year funds. This means that the latter year funding will sit unexpended for months or years. Fourth, our recent visits to certain universities revealed a lack of communication between the grantees (the PIs), their university business offices, and AFOSR. While we have been communicating to the grantees the importance of quickly expending funds, that message does not seem to be getting to the business offices. We recently sent the business offices a copy of expenditure goals and a memorandum reiterating that their lack of timely expenditures could result in funding cuts to their grants. Many of the business offices were not meeting with their grantees on a regular basis to review the execution of grants, and we believe our continuing efforts to heighten business offices' awareness of the issue will help correct some of the problem.

Finally, part of the problem in meeting expenditure goals has been self-inflicted by DoD. The current goals followed by the Office of the Secretary of Defense (OSD) are not appropriate for a basic research program. They tend to be based on operations and management (O&M) budgets being funded beginning on October 1 of each fiscal year. While meeting with the universities, they informed us that getting their grant funding later in the year causes significant problems for them, including their inability to hire qualified students to conduct research. By the time they received a grant and associated funding in the spring or summer, qualified students were already assigned elsewhere, and the university is forced to delay hiring and research, which delays its ability to invoice AFOSR. If funding was granted earlier in the fiscal year, the problem would be alleviated because the universities could more quickly hire students and start the research. While AFOSR requests adequate funding while under Continuing Resolution authority (CRA), the funding of grants is not considered critical since research is an ongoing process, resulting in AFOSR receiving a fraction of funding requested. In addition, when funding is issued under a CRA, it doesn't come down to AFOSR but rather goes to the MAJCOM and AFRL HQ, both which decide who gets what, creating more opportunities to further erode CRA funding to AFOSR.

The current 6.1 expenditure goals that OSD has set are not achievable for AFOSR, based on the nature of the spending pattern of the basic research program. Instead, AFOSR has worked with AFRL/FM to create new, realistic expenditure goals that OSD should allow AFOSR and AFRL to use. A special exception should be written in the regulations to deal with the grant process in reference to fund execution. In addition, RDT&E funding is considered a 2-year appropriation, yet AFOSR is required to obligate a fiscal year's funds on grants as if they were a 1-year appropriation; this artificial constraint further exacerbates the expenditure problem.

The nature of the University Research Imitative (61103F) program is such that a grant, when issued to a university, many times has that organization partnering with other entities in the conduct of research. The main grantee, the external organization receiving the grant from AFOSR, cannot invoice AFOSR until the other entities it has partnered with invoice them. This trail of invoicing has a tremendous impact on expenditure rates for the extramural program. AFOSR studied the NSF model wherein they issue multiple grants, one to each of the partnering entities in addition to the external organization, thereby significantly improving the expenditure of funds in their extramural program. While somewhat effective and potentially valuable to AFOSR, we believe this would represent a significant increase in workload, roughly 400 percent, with a large increase in contracting personnel.

Forwarding Panel

The panel for the discussion with AFOSR program staff on "forwarding" included Dr. Van Blackwood, principal assistant to the chief scientist; Dr. Tatjana Curcic, program officer, atomic and molecular physics; Dr. Hugh Delong, program officer, natural materials and systems; Ms. Molly Lachance, program analyst, digital outreach, and Dr. Kent Miller, program officer space sciences. The panel addressed the topic of "forwarding" Air Force-relevant research—for example, impact of Open Access requirements, intramural program, STTRs, partnerships with industry, collaboration with DARPA, patent or publication metrics, relationship with Air Force MAJCOMs, and other Air Force stakeholders.

Tatjana Curcic, Program Officer, Atomic and Molecular Physics, AFOSR/RTB

I was on the "Forwarding" panel and gave the example of the cold-atom research at AFRL, for future precision navigation capabilities in GPS-denied environments, its history and the role that AFOSR played in this research program since its inception. Let me know if you would like me to try to summarize that separately. The other comment that I remember clearly is my response to Dr. Godfrey's question at the end, and here is how I would summarize it (with some additional comments): (1) AFOSR has always been a lean organization, but today it's bare bones, and there is no end in sight to the worrisome trend (example: recent significant reduction of IT support). AFOSR's operational costs (i.e., overhead) were at about 10% of the total budget four years ago, and today that number is down to 6.5 percent. It is perhaps enlightening to put that in perspective by comparing with similar organizations. What I have recently learned (unverified) is that ONR is at about 9-10 percent, ARO at over 20 percent, HSARPA at around 20 percent, and DARPA at about 15 percent. It might be interesting to look at NSF and DOE's Office of Science as well. What about other AFRL TDs? (2) We are all feeling the squeeze very much. AFOSR POs spend much of their time doing administrative work that not only does not require a PhD, but could be accomplished, probably more effectively, by a college grad or even a smart high-school grad. More importantly, this is the time taken away from the part of the mission that only our highly trained (and expensive) experts (POs) can do—read and evaluate scientific literature and interact with the research community, so that we can best identify future directions for our programs and evaluate present investment and thus pave the way for critical technology developments for the USAF in the future, much like what we did for the cold-atom-based precision navigation technology. This is not a good use of Air Force resources.

Kent Miller, Program Officer, Space Sciences, AFOSR/RTB

I agree with the comments from the others, but would like to add one other observation. I am feeling a pull from the laboratories toward applied research under the instructions to become "more relevant." I thought that was also true in the questions to my panel on transitions. It was probably to a large part because of a specific example that a member of the Studies Board was discussing with a panel member. There are probably special cases where AFOSR is able to respond to rapid response requests, but that is the exception. In most of our portfolios, if we are positioning our research for immediate relevance or "rapid response," we are dealing with evolutionary research, not revolutionary.

Former AFOSR Directors Panel

The panel of former AFOSR directors consisted of all but one of the former AFOSR directors who served during the past two decades, plus one of the acting directors.

Joseph Janni

The Air Force Office of Scientific Research (AFOSR) has an exemplary track record of discovering and selecting the most promising basic research ideas from Universities, the Air Force laboratory system, and industry. AFOSR then sponsors the best ideas with the greatest likelihood of benefitting the AF of tomorrow. This process is guided by the long-term needs of the AF. The approach of actively soliciting, evaluating, and funding scientific breakthroughs well before they are mainstream research areas has resulted in technologies that have proven to be of great benefit to the AF.

The complete list of AFOSR achievements is much too long to delineate here.

From its' inception, the Air Force has been a highly technical organization, dependent upon scientific and technical advances to remain dominant first in the air domain, then space, and recently cyberspace. This need for technical dominance led to the establishment of AFOSR over 65 years ago, with great success. AFOSR is uniquely capable of understanding and pursuing long-term scientific trends as proven by the historical record. Degrading this capability, as was attempted last year, would diminish the long term health and capability of the Air Force.

AFOSR must be strengthened and solidly supported by senior AF leadership. Efforts to reorient basic research funding toward more applied activities would be a serious and shortsighted mistake. Mediocre management approaches and weak administrative methods must not be implemented under the guise of efficiency, nor arbitrary reorganizations imposed that would have unintended deleterious results.

The success of Air Force basic research depends on the high-level education, professional background, meticulous methodology, and insightful approaches of the organizational leadership and POs (previously program managers) in selecting the most promising basic research proposals. These are necessary talents that AFOSR POs have always had and that AFOSR's senior leadership must retain the authority to exercise. Recent drastic reductions in the number of POs are degrading capabilities and must be reversed.

An imbedded contracting office is crucial to success. AFOSR is the basic research manager of the Air Force, carrying out long-term research by informing universities, industry, and the AFRL of long-term requirements via broad-agency announcements, intra-laboratory communications, and conferences. Research proposals are then submitted to AFOSR and evaluated by AFOSR POs using accepted methodology. When approved, the proposals are sent to the AFOSR contracting office for prompt award. It is absolutely critical that the POs and contracting personnel work closely together in a timely manner. This process requires contracting personnel that are co-located with the POs and a contracting office imbedded with AFOSR. Long distance procurement would not work.

Management of contracting was removed from the direct authority of AFOSR's director in 2007, diluting the concept of the imbedded contracting office. This function now reports to the AFRL Contracting Functional instead of the AFOSR director. The rationale should be reassessed and line authority returned to the director. These comments also apply to finance and personnel.

The single manager method has been very successful. It enables ready implementation of OSD and Air Force guidance, and provides focus, and cohesive trend management across the Air Force.

Some unvarnished history. During my tenure as AFOSR Director the teamwork between all Chief Scientists, technical personnel, and the AFRL Commander was excellent. Sadly that was not the case with the "support" organizations at WPAFB such as Plans and Programs, Finance, and Contracting. They were uncooperative, intransigent, and relentless in their efforts to diminish and degrade AFOSR. It was blatant parochialism and an attempt to acquire more power and funds. That attitude seems to still exist but must be changed to be more supportive, constructive, and positive.

Lyle H. Schwartz

Collectively, broad technical program areas selected for emphasis in 6.1 must reflect both opportunity and need. This selection process requires high-level strategic planning, especially by the AFOSR director in collaboration with chief scientists from throughout AFRL, and demands personnel of strong but wide ranging background.

Individuals who are at the heart of specific program topic selection, the POs, are expected to *build* as well as manage programs. POs who first spend some time in one of the DoD laboratories have a leg up in the area of understanding the mission aspect of their work. Their skillset is often supported by POs with university background (either permanent transition or IPA visitor status) and those who have spent some time in industrial settings. While it is increasingly difficult to cover all technical areas, those areas selected must still be managed by technical experts with vision and a continually updated knowledge base. Regrettably, restrictions on travel opportunity and reduced program assistant support within AFOSR has led to less interaction with laboratory scientists and the rest of the extramural research community.

A critical component of the execution process is the need for a strong 6.1 program within the laboratory directorates. Only within the laboratory is there the talent, flexibility, and organization required to interact with the university community (the source of much forefront research) and with the developers and ultimate users of the technology products. This process is facilitated when the intramural and extramural programs are viewed as elements in an integrated program overseen by technical POs from AFOSR.

Patrick Carrick

There have been efforts of the Air Force and DoD in the last 4 years to become "more efficient" by shifting resources from people and support into the "mission." The efforts of AFOSR to become a "more efficient" organization predated this Air Force/DoD push, with significant results for the optimization of AFOSR as the premier "find, fund, forward" science and technology organization. This increased optimization was driven by the transition from paper into electronic program documentation, an electronic tracking system to optimize approval and funding of grants and contracts, and improved IT and communication processes to interact with the Air Force and overall national and international scientific communities. However, when a lean, efficient organization is forced to "become more lean" by a parent organization, the result is actually reduction in overall efficiency. This is evidenced by the lack of hiring authority, reduction in UMD slots, loss of hiring authority for IPAs, and severe restriction of contractor support that AFOSR experienced over the last 5 years. Lack of ability to hire new program managers with superior experience, vision, and connection to their scientific community will lead to degraded decisions on their part about new scientific trends, poor funding decisions, and decreasing respect of the scientific community and DoD. In addition to considerable degradation in program manager hiring and reduced quality, the loss of senior executive oversight of the vast majority of the scientific and technical decision making processes has led to degraded interaction at the same senior level with like-organizations within DoD and the federal government. Lack of senior-level engagement at a technical level will ultimately jeopardize the ability of AFOSR to coordinate strategy within the Air Force, DoD, and other federal agencies and degrade high-level funding decisions about the future direction of Air Force capabilities.

In addition, efforts to become "more lean" and efficient in supervisory chains led to considerable pressure to make AFOSR support organizations (contracting, finance, human resources) report directly to (or actually stationed at) the parent organization more than 450 miles away. Having a highly integrated and self-reliant organization separate out key functional and executing groups (contracting, finance) from the decision and action groups (leadership and program managers) will lead to a less efficient organization. It is especially important to keep a close, engaged contracting and finance shop completely integrated with the program managers and AFOSR leadership.

Current trends in acceleration of scientific and technical advancement worldwide will likely to continue for some time. This makes the ability to engage and cooperate with international colleagues increasingly important. AFOSR is strongly positioned in global engagement, with offices in three international locations. In addition, international cooperation has a political aspect. AFOSR is uniquely positioned within the Air Force for across-the-board cooperation with nearly any country, as issuance of grants and cooperative programs in basic research do not require specific country-to-country agreements. The location of AFOSR in the Washington, D.C., region also strengthens this interaction, as many embassies in the United States, located in Washington, have liaisons with scientific and technical responsibilities. This combination of ease-of-interaction with ease-of-access to international embassy scientific and technical staff makes AFOSR a valuable commodity for the Air Force.

Tom Russell

Thomas Russell believes that the AFOSR director needs as much autonomy to operate as the POs do. Then Maj Gen Pawlikowski, AFRL/CC, recognized this fact. The reduction in IPAs was related to overall reduction in SESs by Secretary Gates, but this has since been rescinded. Dr. Russell believes that centralization of basic research at AFRL (or ARL) headquarters is not healthy; there needs to be representation in Silicon Valley, Research Triangle, and other centers of innovation. The force of the Air Force, in his opinion, in 2025 has largely already been built; there is very little that AFRL and AFOSR will contribute to this force. To its detriment, according to Russell, the Air Force has become very near-term focused; the Army is actually more long-term focused. In his opinion, there is no real budget issue with respect to 6.1; these are artificially created. The center of science is moving to Asia today. Basic research is heterogeneous (i.e., is different for each Service). With respect to personnel, Russell noted that ARL is using IPAs. AFOSR's work-flow process is the best he has ever seen. Finally, during his tenure, administrative activities comprised 28 percent of a PO's responsibilities.

Brendan Godfrey

AFOSR has a long history of identifying and funding outstanding research opportunities, highlighting important research advances funded by others, transitioning technologies, and building strong ties between academia and the Air Force, particularly AFRL. Establishing ties with young researchers is especially important, and AFOSR does so through its vigorous Young Investigator Program, its National Research Council postdoctoral program, and its extensive support for graduate students. Its international offices in Europe, Asia, and South America tap outstanding research overseas, build effective ties with overseas researchers, and create international goodwill. AFOSR has a collaborative, risk-taking culture. Its automated workflow and information management systems enhance its cost-effectiveness. AFOSR has by far the lowest overhead in AFRL.

Unfortunately, AFOSR's mission performance and reputation recently have been challenged by several issues. The directorship has turned over repeatedly in the past 2 years, and senior leadership positions have been abolished, triggering otherwise needless and disruptive reorganizations. Several well-qualified POs have left AFOSR, in large part due to the recent ill-conceived attempt to move AFOSR to Dayton, Ohio, and, more generally, to declining morale. Numerous vacancies exist, reducing the overall effectiveness of grant selection, grant oversight, and technology transition. The highly effective IPA no longer is being used, compounding recruiting difficulties. Moreover, POs are not supported as effectively as in the past due to too few assistant program officers, wasteful travel restrictions, slow human and animal use determinations, and inadequate library access. Because timely contracting plays an important role in AFOSR mission performance, AFOSR contracting should report to the AFOSR director.

Robin Staffin, Director of Basic Research—
Office of the Assistant Secretary of Defense for Research and Engineering

In Staffin's opinion, 6.1, 6.2, 6.3 research do not follow a simple linear model, although they are related. The relationship between the Office of Basic Research that he leads and AFOSR is a close one, given their proximity and relative sizes and missions in basic research. On the other hand, given its geographical distance and its heavy contract management workload, the relationship with AFRL is the more distant one of the Service laboratories. Staffin stated that it is his assessment that AFOSR is doing a good job in identifying and supporting transformative areas in basic research- but he wishes they had more program officers and assistant program officers. Further, Staffin is satisfied, and actually impressed with the scientific quality and expertise of the people working in AFOSR. Staffin said it was his impression that AFRL does not adequately appreciate the value that AFOSR brings to DOD science, or the caliber of its program officers, the importance of their work for the Air Force, and that more are needed. He noted that DOE laboratories have a widely recognized scientific legacy, that include such leaders as Oppenheimer, Lawrence, Teller, Panofsky, and Wilson; DoD laboratories do not emphasize their scientific legacy, and that he believes is a loss to many of their personnel. He would like to better understand the metric of success in DoD laboratories. Staffin noted his experience of DoD laboratories as being relatively stand-offish compared to other agency laboratories; and noted the irony that they come across to some in academia as more ivory tower than universities. Staffin reflected on the importance of laboratories in their being able to reach outside DoD and the United States, to maintain their scientific and engineering competitiveness.

Christopher Thomas (SES), Administrator—Defense Technical Information Service

Christopher Thomas spoke on public access impact on basic research, requirements, and benefits, and how DTIC's products and services will change through making public access journals and data sets available to the public. DTIC's mission is to provide essential technical RDT&E information rapidly, accurately, and reliably to support our DoD customers' needs. Under DoDI 3200.12, *DoD Scientific and Technical Information Program (August 22, 2013)*, and DoDM 3200.14, *Principles and Operational Parameters of the DoD Scientific and Technical Information Program (STIP) (March 14, 2014)*, DTIC collects Scientific and Technical Information (STI) and disseminates it for the benefit of the DoD community, industry, and academia, to encourage the reuse of information. The White House Office of Science and Technology Policy (OSTP) memo "Increasing Access to the Results of Federally Funded Scientific Research" dated February 22, 2013, requires all federal agencies with over $100 million in annual research and development (R&D) expenditures to support increased public access to the results of research funded by the federal government, including peer-reviewed scholarly publications and digitally formatted scientific data arising from unclassified unlimited research necessary to validate the results and conclusions of the effort (with exceptions for national security, proprietary data, and data that would be more expensive to retain than it is worth).

The overarching goal is to accelerate scientific breakthroughs and innovation, promote entrepreneurship, and enhance economic growth and job creation. The benefit to DoD is increased access to government-wide scholarly publications and scientific data and cost savings to DoD organizations (no need to buy back DoD-funded content); industry will have increased visibility of DoD research priorities which provides the opportunity for industry to invest in DoD areas of interest. Under Secretary for Acquisition, Technology and Logistics, Mr. Frank Kendall, signed a memo on July 9, 2014, conveying support to the White House mandate and directing DoD compliance. The Defense Basic Research Advisory Group (DBRAG) is creating a directive-type memo to require public access (data management plan at the start of every effort, publications, and metadata and location of data sets submitted to DTIC) for intramural basic research, starting January 2016. Extramural research, which will require regulatory changes for contracts and grants, will come into effect January 2017. Then, in 2018 and 2019, public access requirements will expand to other areas besides basic research. DTIC is working with other agencies such as the National Institutes of Health (NIH), NSF, and DOE to use best practices. The DoD plan is at http://dtic.mil/dtic/pdf/DoD_PublicAccessPlan_Feb2015.pdf. The group at the workshop was very interested and discussed the need for a federated search of data sets, how to handle software associated with data sets, and the size of some of the data sets (such as wind tunnel data). Also discussed were the benefits of gathering public access articles in terms of providing POs with situational awareness of the field and analyzing the research being done in the department overall.

Christopher Fall, Assistant Director for Defense Programs—Office of Science and Technology Policy, Executive Office of the President

Christopher Fall noted that the new *National Security Strategy* was published in February 2015 and that a corresponding S&T strategy is in development.[1] OSTP's role is not to set the science priorities for federal agencies, but to facilitate the process. Fall then discussed the legacy national security S&T enterprise challenges, including constrained budgets, workforce demographics, aging facilities, outdated processes, and governance. Falls observed that DoD is out-in-front on international research, but it is not clear how well this effort is coordinated. He noted that DoD's Small Business Innovation Research (SBIR)/STTR programs are well run and suggested that DoD basic research grants include "broader impact" criteria.

Air Force Chief Scientist Panel

The panel of chief scientists from the Air Force major commands included four current and former Air Force command chief scientists. Panel members were asked to provide their thoughts, as end users, on the effectiveness of AFOSR's mechanisms to stay abreast of emerging Air Force technology requirements (PULL) and ability to establish transition pathways for research that has evolved beyond the scope of basic science (PUSH).

---

[1] Executive Office of the President, *National Security Strategy,* February 2015, https://www.whitehouse.gov/sites/default/files/docs/2015_national_security_strategy.pdf, accessed May 26, 2015.

Azar S. Ali, Chief Scientist—Pacific Air Forces

Pacific Air Forces (PACAF) has a large area of responsibility (AOR) that covers 10 time zones, 36 nations with 70 percent of the world's population, over 1,000 languages, and over 50 percent of Earth's surface. Seven of the ten largest standing militaries in the world are in PACAF's AOR. Some of these militaries are very capable and have developed anti-access area denial strategies. DoD has recognized the many challenges facing PACAF and has made a deliberate effort to 'Pivot to the Pacific.' The PACAF Chief Scientist's Office is hosting a Science and Technology Information Exchange, June 15-19, 2015, at Hickam Air Force Base, Hawaii. The intent is to engage the S&T community and the tremendous talent and resources of AFOSR, and in particular, to address some of the challenges facing PACAF. Three such challenges are the need for (1) Rapidly Converging Algorithms for Swarming Weapons Engaging Mobile Targets, (2) Constrained Algorithms for Apparent Random Tracks for Swarming Munitions Engaging an Integrated Air Defense System, and (3) Constrained Optimization Algorithms to Enable Logistics Flow in Support of Flexible Basing.

Janet S. Fender, Chief Scientist—Air Combat Command

Dr. Joseph Janni profoundly changed the culture of AFOSR, sharply focusing the organization on emerging Air Force technical requirements and establishing technology transition pathways. In fact, Dr. Janni set a precedent for directly transitioning 6.1 research to operational capabilities when AFOSR supported decision aids for the Airborne Laser program and when AFOSR research enhanced operational capabilities at the Maui Space Surveillance Site. Continuing that trend, AFOSR today directly supports the Air Force Weather Agency, now a wing in Air Combat Command (ACC). AFOSR also supports a post-doctoral research on flexible electronics for ACC/SG medical applications as well as ACC/PR personnel recovery (Combat Search and Rescue). AFOSR could energize innovation by focusing multi-disciplinary research on grand challenges for the Air Force. For example, enabling low SWaP, ultra-precision, stable timing would be a landmark accomplishment for AFOSR and revolutionize Air Force operations. Expanding the AFOSR rotational assignment program to offer tailored experiences at operational locations would certainly "blue" AFOSR program managers. Bringing transition partner presentations into AFOSR workshops could formalize pathways and expedite delivery of integrated capabilities by AFRL. AFOSR can play an important role in integrating research performed by multiple AFRL directorates—especially if focused on a grand challenge or enabling integrated capabilities.

Donald R. Erbschloe, Chief Scientist—Air Mobility Command

At first glance, the relationship between the basic research community, which explores new concepts and fundamental understanding, and the warfighter, who depends on proven, reliable, fielded technologies for mission execution, would seem to be, at best, one of distant, nodding acquaintance. However, AFOSR has a long and successful track record of engagement with the Air Force MAJCOMS. In Air Mobility Command the last two decades have seen productive partnerships in such areas as precision airdrop, fuel efficiency, algorithms for dynamic optimization, mitigation of radar clutter, and access to top international science. Basic research is the bedrock for Air Force science and technology. AFOSR informs the MAJCOMs on what is possible, and the MAJCOMs communicate needs and areas for potential applications.

J. Douglas Beason, Former Chief Scientist—Air Force Space Command

AFOSR plays an important role in insuring that warfighters have access to the world's best weapons needed to win the fight. As combat becomes more sophisticated, new weapons' capability increasingly depends on exploiting advanced technology. This advanced technology does not spring out of thin air or occur because some administrator says to "make it happen"; rather, advances in technology ultimately depend on investments made in basic research, and usually in areas focused on defense. But the traditional avenues for supporting and obtaining Air Force-focused basic research results are dwindling: the vast majority of commercial and industrial R&D is concentrated on short-term, time-to-market products, and the infrastructure needed to conduct R&D outside of their market may not be sustained. Defense-oriented basic research in academia and government laboratories may be focused in areas not applicable to the Air Force—for example, extending pit or nuclear waste container lifetimes for DOE. And allocating basic research monies to individual AFRL directorates will result in focused, insular investments. As such, having AFOSR as a separate agency with a strategic knowledge of Air Force warfighting needs continues to be the optimum way of making basic research investments.

SESSION 2: MAY 27-28, 2015

There were nine confirmed speakers for the second workshop session. Listed below, in chronological order, are short abstracts of the speakers' remarks.

Mica R. Endsley—Air Force Chief Scientist

AFOSR is unique in its focus on 6.1 basic research for the Air Force. With some 70 Nobel Laureates included in its past award recipients, it has been highly successful in this mission. I believe that the current structure of AFOSR, as a separate directorate within AFRL, is effective in preserving its ability to focus on basic research without the pressure to slide towards more near-term directed research.

My expectations of AFOSR are first that it support basic research that will enable the Air Force to achieve new goals along its core missions, expand its capabilities, and address new challenges associated with activities in air, space, and cyber. These activities should be closely aligned with long-term mission needs and challenges as spelled out in our Strategic Vision and the master plans from each of the core function leads. Its second goal is to support basic research that is more bottoms-up, identifying potential new areas of exploration that may be advantageous to the Air Force, even if we have not yet identified it as a need. This is a careful balance between the directed and undirected, which requires extensive knowledge of developing fields and the imagination to see long-term utility in very early R&D. In general, I think AFOSR is meeting that need. However, it is always a good idea to carefully review research in each portfolio to determine what is fruitful and what needs to be redirected on a frequent basis. While AFOSR is strong in certain areas, a balanced portfolio may require more new work in emerging fields (e.g., cyber, autonomy, quantum).

In general, AFOSR works within the AFRL process to determine its research directions and funding expenditures each year. The chief scientist is invited to program reviews and to provide input as requested on particular topic areas. AFOSR has also provided extensive reviews of its international portfolio and operations. The chief scientist provides general support (e.g., advocacy for 6.1 research on the Air Staff) in my role in support of S&T for the Air Force.

By definition, AFOSR should be focused on the long-term basic research needed by the Air Force. It does, however, also need to ensure an effective transition path from successful basic research work to applied research, and advanced technology development is needed to move this work into practice. This can best be accomplished by a close working relationship with AFRL to identify the successful research that needs to be transitioned and creating a funding stream and program management to move that work forward. The Air Force has created a set of long-term challenges associated with contested environments (A2AD), hypersonics, directed energy, autonomy, unmanned systems, nanomaterials, and cyber security and reduced costs for operations across the board. Much of AFOSR's portfolio is directed at these efforts; however, some more focused effort may need to be directed at specific gaps or opportunities in these areas. In addition, I expect that AFOSR may find many new opportunities for advancements that are not on this list, but which may greatly enhance our capabilities.

Morley O. Stone, Chief Technology Officer—Air Force Research Laboratory[2]

AFOSR is an integral part of AFRL. As the sole steward of basic research (6.1) funding in the Air Force, it has a unique and foundational role in the laboratory and the Air Force as a whole. That role is one of science and technology "scout"—that is, monitoring emerging fields in laboratories across the world and harvesting and directing these scientific breakthroughs to the greater Air Force science and technology enterprise. Just as important as its outward-facing role, AFOSR also plays an integral role in supporting intramural research within AFRL. By fostering the development of PIs within AFRL, it allows the laboratory to "grow" world-class researchers internally, creating deep, in-house technical expertise that is called on time and time again to solve the Air Force's most challenging problems.

My comments to the panel will be based on the various capacities in which I have interacted with AFOSR. First, I interacted with AFOSR as a PI—my interactions were almost exclusively at the PI to PO level. Later in my career as the chief scientist of the Human Performance Wing, my interactions were with AFOSR leadership and shifted to more strategic interactions involving the development of human capital (i.e., NRC post-doctoral program and the defining of mutually agreed upon future joint technical ventures, such as, neuroscience and synthetic biology).

---

[2] Abstract cleared for public release, case number: 88ABW-2015-3216.

Finally, I will conclude my comments reflecting on the AFOSR of today in comparison to my reflections of the AFOSR I perceived early in my career. Certainly, changes have occurred, and in my opinion, many of them have been detrimental to the mission of AFOSR—i.e., the best way to prevent technical surprise is to create it. In my reflection and resulting analysis, relatively minor changes could be made to begin the restoration of AFOSR as a key player in this nation's S&T enterprise. These changes include greater use of IPA authority, the authorization of more senior level positions (SES and ST), greater rotation of POs, more POs to adequately cover emerging areas, mechanisms to foster greater portfolio turnover every year, and the proper, *agile* support of the AFOSR mission by the supporting functional offices.

Rita Colwell, Distinguished University Professor, University of Maryland, Former NSF Director

Dr. Rita Colwell served as NSF director from 1998 to 2004. Up until 1998, NSF worked independently, but had relationships with ONR, the Coast Guard, and NASA. The interactions with the Air Force were not as pronounced. 9/11 was a landmark. NSF did not and does not do classified research, but Dr. Colwell established relationships with DoD agencies so that DoD personnel could attend NSF meetings. NSF could serve as a facilitator between researchers and other agencies. How did NSF approach management of basic research? The answer is, all NSF research is basic research. NSF can issue grants up to $200,000 on a rapid basis, if needed. There are cases where NSF sponsors basic research directed to a national need. Rotators serve 3-year appointments. The peer review process is the jewel of NSF. Dr. Colwell instituted a rule that 5 percent of a project manager's research portfolio had to be "risky" research. Transparency with this process is very important. NSF can support scientific conferences. At this moment in time, there is a partnership between NSF and AFOSR—$5 million to $10 million per year. Dr. Colwell cited an example of a cooperative weather satellite program. Fundamentally, the NSF-AFOSR relationship is science-driven. The Air Force contributes to the NSF Research Experience for Undergraduates program. NSF's Math and Physical Sciences directorate interacts with the Air Force in operating the Sacramento Peak observatory. NSF's relationship with AFOSR is extensive, productive, and should be continued. It is important to set research priorities. Dr. Colwell would recommend to AFOSR peer review, transparency, and communication with its constituencies, including the general public. Good management is a central component of the NSF portfolio, and NSF is recognized by the Office of Management and Budget as the best-managed federal agency. Dr. Colwell noted that NSF research areas could align more effectively with Air Force research areas.

David Skatrud (SES), Director—Army Research Office

The ARO utilizes a number of formal and informal procedures and processes to lead the discovery of innovative scientific breakthroughs that significantly advance technologies for critical new Army warfighting capabilities. Program managers are encouraged and resourced to fund highly innovative, high-risk extramural basic research (mostly with universities, but also with some small and large companies), which has potentially disruptive and game-changing Army operational implications. Scientific areas of interest are identified by consideration of Army near- and long-term strategic and tactical requirements, in addition to determination of incipient scientific advances that can impact those requirements. Proposed projects are subject to a pre-selection review process that addresses both scientific quality and Army impact. Program managers are given autonomy to select high-risk projects but are held accountable through biennial reviews by independent review boards that access the overall scientific quality and Army relevance of the program manager's research portfolios. In addition, technology transition is affected through involvement of stakeholders from the Army and DoD R&D community in all phases of the research: program formulation, project selection, and program execution.

Linda G. Blevins, Ph.D., Senior Technical Advisor— Office of the Deputy Director for Science Programs, Office of Science, Department of Energy

The mission of the DOE Office of Science is to deliver the scientific discoveries and major scientific tools that transform our understanding of nature and advance the energy, economic, and national security of the United States. The office is the nation's largest federal sponsor of basic research in the physical sciences and the lead federal agency supporting fundamental scientific research for energy. The Office of Science budget of about $5 billion per year supports research in six program offices: (1) Advanced Scientific Computing Research, (2) Basic Energy Sciences, (3) Biological and Environmental Research, (4) Fusion Energy Sciences, (5) High Energy Physics, and (6) Nuclear Physics. About half of the budget supports operation of scientific user facilities and construction of new facilities; the other half supports research at national laboratories and universities. The Office of Science directly supports about 25,000 Ph.D.s, postdoctoral researchers, graduate students, and undergraduate students per year. Our scientific user facilities host another 29,000 users per year.

The Office of Science program staff members, who work primarily in Germantown, Maryland, have various ongoing interactions with AFOSR staff. Interactions include but are not limited to attending each other's PI meetings; serving as reviewers of each other's proposals and programs; co-chairing interagency working groups such as the Networking and Information Technology Research and Development (NITRD) Program; co-organizing federal funders' sessions at scientific conferences; and issuing at least one joint solicitation. The two organizations held a high-level research summit hosted by Brendan Godfrey and Patricia Dehmer in 2009 to exchange program information and discuss mutual interests. Interactions currently take place in the scientific areas of software-defined networking, quantum computing, cybersecurity, catalysis, computational and theoretical chemistry, solar photochemistry, atomic and molecular physics, ultrashort pulse laser-matter interactions, plasma and electro-energetic physics, particle and radiation sources and their applications, and laser technology development.

The Office of Science employs a variety of methods in managing its programs that might be of interest to AFOSR. These include varying the research funding modalities to include smaller single investigator grants, mid-size Energy Frontier Research Center grants, and larger Bioenergy Research Centers and Energy Innovation hubs. The Office of Science is a steward for 10 of the DOE's 17 national laboratories, specializing in design, construction, and operation of a large suite of scientific user facilities available without user fees. The office hosts basic research needs and other types of workshops and works closely with its six federal advisory committees. We coordinate with other agencies to minimize researcher burden. Every 3 years, each of our programs is critically reviewed by a Committee of Visitors. The Office of Science employs about 150 expert scientific program managers who execute peer review as the cornerstone of their work. We have been, over the past few years, harmonizing business practices through the design and implementation of new, end-to-end business software, the Portfolio Analysis and Management System (PAMS). We have also begun to implement common programs across the Office of Science such as the Early Career Research Program.

DOE in February 2015 announced the formation of a new Office of Technology Transitions to help expand the commercial impact of DOE research. The acting director of this new office is Jetta Wong. The office is working closely with the national laboratories and to engage with industry to commercialize technology and strengthen the global competitiveness of U.S. industries based on scientific and technological innovations.

Walter Jones (SES), Executive Director—Office of Naval Research

Although the three Service S&T organizations are structured very differently, if you have the right people, the organizational model is not that important. For the Department of the Navy, ONR is the single manager of all S&T. ONR does not perform any in-house S&T, but funds others to do work, including universities, industry, and government labs. ONR is an integrated S&T organization (basic research, applied research, and advanced technology development). Melvin asked the question, how does ONR define basic research? A: We use the standard DoD definition. The Chief of Naval Research (2-star admiral) reports to SECNAV thru Assistant Secretary of the Navy (Sean Stackley). ONR funds less than 50 percent of NRL's work, the remainder is customer funded. It is really hard to find uncommitted 6.4-6.7 $$ to fund transitions from S&T to programs of record. There are six S&T departments within ONR, each one led by a Tier-2 SES. There are also two SES directors (Research and Technology). Largest portion (about 45 percent) of ONR funding is for basic and early applied research. ONR is responsible for all Future Naval Capability efforts (each S&T program is 2-4 years long), with a 70-percent transition success rate. About 15-20 FNC programs started each year, for those programs that are not transitioned the reason is typically that customer transition funding falls through late in the project. DoD Reliance 21 program coordinates S&T across the Department of Defense (17 Communities of Interest). Conference attendance approval process is very time-consuming, with conference expenses $100 K and above having to be approved by SECNAV (used to be $500K threshold). ONR has six overseas offices and shares three locations with AFOSR (London, Tokyo, and Santiago). There are eight Tier-1 SES positions within the S&T Departments at ONR - we are currently trying to fill four of these slots.

Robie Samanta Roy, Vice President, Technology and Innovation Lockheed Martin Corporation

The topic of congressional involvement with basic research institutions like AFOSR focuses primarily on overall funding levels and secondarily on topical research areas. At a practical level, basic research is a funding mechanism for universities in congressional home districts. Many members may have specific interests with their individual academic institutions and various research areas. From a DoD perspective, a desirable situation is one where members' interests align with DoD's interests and needs. There are situations where relevant committee staff may recommend to members against certain special interests if there is not a strong alignment with DoD or if proposed funding on certain topical areas will lead to inefficient use of limited resources. To ensure a higher probability of congressional action, it is important that the authorizing staff be linked to the appropriations staff in both the Senate and the House. Generally speaking, congressional support for basic research is positive when basic research can be linked ultimately to successful warfighter support.

University Panel Discussion

The University Discussion Panel included Dr. Denis Wirtz, vice provost of research, Johns Hopkins University; Dr. Bill Bonvillian, MIT Washington, D.C. Office; and Dr. William Melvin, Director, Sensors and Intelligent Systems Directorate, Georgia Tech Research Institute, speaking for Dr. Stephen E. Cross, Executive Vice President for Research, Georgia Institute of Technology.

Dennis Wirtz, Vice Provost of Research, Johns Hopkins University

Johns Hopkins University (JHU) is the only school with over $ 1 billion ($1.7 billion) in annual research funding. Concerns about increasing administrative burdens on young researchers and may drive away researchers. We are not that good at landing the larger center grants. Until recently, we could count on faculty to be competitive on their own in securing funding without institutional assistance. Within the last 5 years, faculty have had to spend more time writing proposals. We think there is a research "valley of death," and we try to assist faculty in navigating through this. JHU has a relatively small number of grants from AFOSR; it has always been relatively easier to obtain grants from NIH and NSF, and so did not have to make efforts to approach other government agencies for grants. This is starting to change. JHU is trying to expose faculty more and more to information sessions to demystify organizations like DARPA. There are avenues and areas where DoD agencies can team together with academia (joint programs).

Bill Bonvillian, Director, Massachusetts Institute of Technology, Washington, D.C. Office

One significant research management question is how to better build strong basic research efforts into applied projects. Advanced manufacturing is the federal government's major new applied R&D initiative of the past 5 years, now being funded at over $500 million per year, with bipartisan support. DoD has been the leading agency involved because of the department's profound stake in U.S. manufacturing capability. After briefly noting core lessons about the nature of the U.S. manufacturing challenge, the discussion turned to new R&D system models designed to deal with it. The main focus of the initiative has been on creating new collaborative manufacturing institutes on a German Fraunhofer model, to perform a technology testing and de-risking role. But attention is now shifting to creating ongoing industry-government-university technology strategies for new advanced manufacturing technology paradigms. And the focus is also now growing on how to integrate research at four agencies, including DoD, into this system of institutes and collaborative strategies. This could present an interesting new approach on how this problem of better integrating a strong research element into an applied project model. Building a cross-agency research collaboration enables leveraging of agency research investments, and avoids the "stranded technology" issues that could face the institutes, because they will be linked to an R&D system to sustain an ongoing flow of technology ideas. The industry-university- government collaboration model, the collaborative technology strategy, and the cross-agency R&D models being developed in advanced manufacturing could inform other areas, including the new DoD Defense Innovation Initiative.

Stephen E. Cross, Executive Vice President for Research, Georgia Institute of Technology

The Georgia Institute of Technology has a long-standing and valued relationship with AFOSR. As the third (or possibly now second)-highest funded AFOSR university, Georgia Tech currently has 45 active, competitively selected projects with $45 million in total funding spanning research in aerospace, information science and technology, materials, sensors, and system technologies. Research at Georgia Tech is guided by a strategy that combines thought leadership, collaborative partnerships, and translation impact. Research in core areas is use-inspired and focused on grand challenges. Shared equipment facilities and administrative services support reduction in cost and administrative burden. There is a long-standing culture between departments to support interdisciplinary research in venues for concurrent discovery, application, and deployment. Special attention is paid to mentoring of junior faculty, and seed grants are used to help them grow new research programs in areas of strategic importance. Georgia Tech actively seeks embedded sponsor presence. In the past 3 years, 10 Fortune 500 companies have established innovation centers on the campus. Many of the practices pursued at Georgia Tech should be applicable to AFOSR, including continued assurance of relevancy of research to critical Air Force needs, reduction of administrative burden to PIs as well as AFOSR program managers, a greater focus on young investigators with mentoring support for "ease of application," and creative means to facilitate PI and program manager engagement with operational personnel (e.g., workshops and conferences on or near operational bases). Endorsement of the recent NSF report on reduction of investigator administrative burden is encouraged, as is the continued presence of AFOSR in the D.C. area.

References

Cross, S., "Georgia Tech's Strategy for Research and Economic Development," Georgia Institute of Technology, Atlanta, Ga., November 16, 2011, http://www.research.gatech.edu/sites/research.gatech.edu/files/Steve%20 Cross%20SR68%2011.16.11.pdf.

Defense Science Board, *Technology and Innovation Enablers for Superiority in 2030*, Office of the Under Secretary of Defense for Acquisition, Technology, and Logistics, Washington, D.C., October 2013, http://www.acq.osd. mil/dsb/reports/DSB2030.pdf.

Murray, W., *Military Adaptation in War*, Institute for Defense Analysis, Alexandria, Va., June 2009. http://www.au.af. mil/au/awc/awcgate/dod/ona_murray_adapt_in_war.pdf.

National Research Council, *Basic Research in Information Science and Technology for Air Force Needs*, The National Academies Press, Washington, D.C., 2006.

National Science Board, *Reducing Investigator's Administrative Workload for Federally Sponsored Research*, NSB-14-18, National Science Foundation, Arlington, Va., March 10, 2014, http://www.nsf.gov/pubs/2014/nsb1418/ nsb1418.pdf.

Perry, J.D., Air Corps experimentation in the interwar years—A case study, *Joint Force Quarterly* Summer, pp. 42-50, 1999.

Pierce, T., *Warfighting and Disruptive Technologies*, Frank Cass Publishing, New York, N.Y., 2004.

Stokes, D., *Pasteur's Quadrant: Basic Science and Technological Innovation*, Brookings Institute Press, Washington, D.C., 1997.

U.S. Air Force Scientific Advisory Board, System-Level Experimentation in Air Force S&T Programs, 2006.

Appendixes

A

Terms of Reference

An ad hoc committee will plan and convene one workshop consisting of two meetings (spaced a month apart for logistical reasons) to identify effective and efficient business practices for the management of AFOSR's portfolio. The workshop will:

1. Explore the unique drivers associated with management of a 6.1 basic research portfolio in DoD and establish the current AFOSR baseline business practices across all its functional offices.
2. Review Army, Navy, and OSD practices for management of basic research that could be benchmarked by AFOSR for incorporation to enhance its own practices.
3. Facilitate a discussion with AFOSR stakeholders (Air Force, OSD, Office of Science and Technology Policy, Congress) as to current and future practices that may further the effective and efficient management of 6.1 basic research on behalf of the Air Force.

The committee will develop the agenda for the workshop, select and invite speakers and discussants and moderate the discussions. In organizing the workshop, the committee might also consider additional topics close to and in line with those mentioned above. The meetings will use a mix of individual presentations, panels, breakout discussions, and question-and-answer sessions to develop an understanding of the relevant issues. Key stakeholders would be identified and invited to participate. One committee-authored workshop report will be prepared in accordance with institutional guidelines.

B

Biographical Sketches of Committee Members

LARRY D. WELCH, *Chair*, is a senior fellow at the Institute for Defense Analyses (IDA) since 2009. He is also a past president and CEO of IDA. Prior to retiring from the U.S. Air Force as a general in 1990, after a 40-year career in the Air Force, he served as follows: from 1986 to 1990, 12th Chief of Staff; from 1985 to 1986 as Commander in Chief, Strategic Air Command; from 1984 to 1985 as Vice Chief of Staff; from 1982 to 1984 as a Deputy Chief of Staff, Programs and Resources; and from 1981 to 1982 as Commander of Air Force Central Command. General Welch's current affiliations include the Air Force Academy Foundation (director); Air Force Space Command Independent Strategic Advisory Group (chairman); Atlantic Council (councilor); Council on Foreign Relations (member); and Defense Policy Board (member), among others. He chaired the National Research Council (NRC) Committee on Department of Defense Basic Research. General Welch holds an M.S. in international relations from George Washington University and graduated from the National War College and the Armed Forces Staff College.

RITA R. COLWELL is a distinguished university professor both at the University of Maryland, College Park, and at Johns Hopkins University's Bloomberg School of Public Health. She is also senior advisor and chairman emeritus at Canon US Life Sciences, Inc., and president and chairman of CosmosID, Inc. Her interests are focused on global infectious disease, water, and health. She is currently developing an international network to address emerging infectious diseases and water issues, including safe drinking water for both the developed and developing world, in collaboration with Safe Water Network, headquartered in New York City. Dr. Colwell served as the 11th director of the National Science Foundation (NSF), 1998-2004. In her capacity as NSF director, she served as co-chair of the Committee on Science of the National Science and Technology Council. Dr. Colwell is a member of the National Academy of Sciences, holds a Ph.D. in oceanography from the University of Washington, and was a member of the NRC Committee on Review of Specialized Degree-Granting Graduate Programs of the DoD in Science, Technology, Engineering, Mathematics (STEM) and Management.

BLAISE J. DURANTE retired from the U.S. Air Force as the Deputy Assistant Secretary for Acquisition Integration, Office of the Assistant Secretary of the Air Force for Acquisition, in Washington, D.C. Mr. Durante managed the acquisition staff organization charged with planning, managing, and analyzing the Air Force's research and development, and acquisition investment budget. He oversaw the integration of research, development, and acquisition budget formulation and execution and directed streamlined management team activities, including Air Force acquisition reform and reduction in total ownership cost efforts. He directed the development of weapon system acquisition policy, including program direction. Mr. Durante served as the chief financial officer for the modernization accounts. As director for Air Force Contracted Advisory and Assistance Services, he directed and was accountable for the Air Force's CAAS programs. He was responsible for acquisition professional development, including directing, coordinating, and reviewing actions mandated by the Defense Acquisition Workforce Improvement Act and Department of Defense (DoD) directives. Mr. Durante also managed acquisition reporting systems and the Air Force's international RD&A programs. He is a member of the NRC Air Force Studies Board.

MELISSA L. FLAGG is a senior program officer at the John D. and Catherine T. MacArthur Foundation where her primary focus is with the fellows program focused on science, technology, engineering, and mathematics. Prior to joining the MacArthur Foundation, Dr. Flagg served as the director for the Office of Technical Intelligence within the Office of the Principal Deputy Assistant Secretary of Defense, Research and Engineering, within the Office of the Secretary of Defense. She also served in roles within the Department of Navy and the Department of State. From 2004-2009, she worked with the Office of Naval Research (ONR) Global, serving first as associate director for S&T policy and force protection in the London office, then as the director of the ONR Global International Liaison Office at headquarters in Arlington, Virginia, and later as ONR's director, strategy and plans. From 2001-2004, Dr. Flagg supported the S&T Adviser to the Secretary of State as a AAAS S&T diplomacy fellow and later as a foreign affairs officer. Her professional awards include the Secretary of Defense Meritorious Civilian Service Award, the Navy Meritorious Civilian Service Award, and the Superior Honor Award from the Department of State. She is a member of the NRC Air Force Studies Board. Dr. Flagg received a Ph.D. in medicinal chemistry from the University of Arizona.

BRENDAN B. GODFREY is a visiting senior research scientist at the University of Maryland, where he conducts studies on numerical simulation of plasmas, participates in committees of the National Academies, and served as advisor to the U.S. Deputy Assistant Secretary of Defense for Research. Previously, he was director of the Air Force Office of Scientific Research (AFOSR), responsible for its nearly half billion dollar basic research program. Known for his contributions to computational plasma theory and applications, he is author of more than 200 publications and reports. He also has served on numerous professional and civic committees. He is a fellow of the IEEE and of the American Physical Society (APS). Dr. Godfrey was a member of the NRC Committee on Review of Specialized Degree-Granting Graduate Programs of the DoD in Science, Technology, Engineering, Mathematics (STEM) and Management and is a member of the Air Force Studies Board. Dr. Godfrey received his Ph.D. from Princeton University.

ZACHARY J. LEMNIOS is vice president, research strategy and worldwide operations, at IBM. In this position, Mr. Lemnios is responsible for the full scope of operations across the twelve IBM global research laboratories. Before joining IBM in 2012, Mr. Lemnios was the Assistant Secretary of Defense for Research and Engineering at DoD, from July 2009 to November 2012. In this position, he provided leadership, strategic vision, and programmatic executive for all near-, mid-, and far-term research and engineering efforts across DoD. He shaped the department's technical strategy to support the President's national security objectives and the Secretary's priorities. Mr. Lemnios served as the chief technology officer at the MIT Lincoln Laboratory from August 2006 to June 2009 where he provided strategic vision for the laboratory as a member of the director's staff and the Laboratory Steering Committee. He holds an M.S. in electrical engineering from Washington University.

WILLIAM MELVIN is deputy director for research at the Georgia Tech Research Institute (GTRI), director of the Sensors and Intelligent Systems Directorate at GTRI, a University System of Georgia regents' researcher, and an adjunct professor in Georgia Tech's Electrical and Computer Engineering Department. His research interests include all aspects of sensor technology development, electronic warfare, computational electromagnetics, signatures, systems engineering/developmental planning, autonomous/intelligent systems and machine learning, threat systems analysis, and quantum science and sensors. He has authored numerous papers in his areas of expertise and holds three U.S. patents on adaptive sensor technology. He is the co-editor of two of the three volumes of the popular *Principles of Modern Radar* book series. Among his distinctions, Dr. Melvin is the recipient of the 2014

IEEE Warren White Award, 2006 IEEE AESS Young Engineer of the Year Award, the 2003 US Air Force Research Laboratory Reservist of the Year Award, and the 2002 US Air Force Materiel Command Engineering and Technical Management Reservist of the Year Award. He was chosen as an IEEE fellow for his contributions to adaptive radar technology and is also a fellow of the Military Sensing Symposium. Also, he is a member of the NRC Board on Army Science and Technology, served on the Air Force Studies Board on Developmental Planning organized through the National Academy of Sciences, and has served on other committees sponsored by the NRC. Dr. Melvin received the Ph.D. in electrical engineering from Lehigh University, as well as the MSEE and BSEE degrees (with high honors) from this same institution, respectively. He is also a distinguished graduate of the Air Force ROTC Program, and a graduate of the U.S. Army Airborne School and the U.S. Air Force Squadron Officer School.

PARVIZ MOIN is the Franklin P. and Caroline M. Johnson Professor of Engineering at Stanford University. He joined the Stanford faculty in 1986. In 1987, he founded the Center for Turbulence Research (CTR), widely recognized as the international focal point for turbulence research. In 2003 he founded the Institute for Computational and Mathematical Engineering at Stanford. He is a co-editor of the *Annual Review of Fluid Mechanics* and associate editor of the *Journal of Computational Physics* and the *Physics of Fluids*. Dr. Moin pioneered the use of direct and large eddy simulation techniques for the study of turbulence physics and control and modeling concepts, and he has written widely on the structure of turbulent shear flows. His research interests include aerodynamic noise and hydro-acoustics, flow control and optimization, large eddy simulation, turbulent combustion, aero-optics, parallel computing, and numerical methods. Dr. Moin is a member of National Academy of Sciences, the National Academy of Engineering, and the American Academy of Arts and Science. He is a fellow of AIAA, APS, and is a current member of the NRC Panel on Mechanical Sciences and Engineering at the Army Research Laboratory. Dr. Moin received his Ph.D. in mathematics and mechanical engineering from Stanford University.

ROBIE SAMANTA ROY is vice president for corporate engineering, technology and operations at Lockheed Martin Corporation. Mr. Roy's responsibilities include leading the corporation's enterprise-level technology innovation strategy to ensure the corporation's continuing ability to develop and leverage new technologies to help solve customers' most challenging problems. In this role, he works with the Engineering and Technology Council and Enterprise Operations leaders to develop and actively manage an enterprise technology roadmap aligned with business area needs, focusing on innovation. He also works with Lockheed Martin's university program with the goal of fostering and transitioning research from leading U.S. research universities, as well as liaison with U.S. government organizations critical to the formation of technical policy and the execution of research. Prior to joining Lockheed Martin, Mr. Roy was a professional staff member with the Senate Armed Services Committee with the portfolio of DoD's wide spectrum of science and technology-related activities. He came to that position from the White House Office of Science and Technology Policy where he was the assistant director for space and aeronautics from 2005 to 2009 and was responsible for space and aeronautics activities ranging from human space flight to the Next Generation Air Transportation System. Before that, he was a strategic analyst at the Congressional Budget Office where he was responsible for studies on military and civil space, missile defense, international relations, and other strategic forces issues. Mr. Roy started his career as a research staff member in the Systems Evaluation Division of IDA from 1995 to 2003 where he conducted studies related to command, control, communications and computers, intelligence, surveillance, and reconnaissance systems. He holds a Ph.D. in aeronautics and astronautics from MIT as well as a master's degree in space policy from the George Washington University and diplomas from the International Space University and Institut d'Etudes Politiques de Paris. Mr. Roy continues to serve in the U.S. Air Force Reserve.

SUBHASH C. SINGHAL is Battelle Fellow Emeritus at Pacific Northwest National Laboratory (PNNL). Dr. Singhal worked as a Battelle fellow and director, Fuel Cells, at PNNL from 2000 to 2012 and provided senior technical, managerial, and commercialization leadership to the laboratory's extensive fuel cell and clean energy programs. Before that, he worked for more than 29 years, initially as a scientist and later as manager of Fuel Cell Technology at the Westinghouse Electric Corporation. While at Westinghouse (which later became part of Siemens), he conducted and/or managed major research, development, and demonstration programs in the field of advanced materials and energy conversion systems, including steam and gas turbines, coal gasification, and fuel cells. From 1984 to 2000, as manager of Fuel Cell Technology there, he was responsible for the development of high temperature solid oxide fuel cells (SOFCs) for stationary power generation. In this role, he led an internationally recognized group in the SOFC technology and brought this technology from a few-watt laboratory curiosity to fully-integrated 200 kW size power generation systems. He has authored 100 scientific publications, edited 17 books, received 13 patents, and given 315 plenary, keynote, and other invited presentations worldwide. Dr. Singhal is also an adjunct professor in the Department of Materials Science and Engineering at the University of Utah and a visiting professor at the China University of Mining and Technology, Beijing, and the Kyushu University, Japan. Dr. Singhal is a member of the National Academy of Engineering and has served on a variety of NRC activities over a span of 30 years. He is a current member of the Board on Higher Education and the Workforce. Dr. Singhal holds a Ph.D. in materials science and engineering from the University of Pennsylvania and an M.B.A. from the University of Pittsburgh.

C

Workshop Sessions and Speakers

SESSION 1
APRIL 27- 28, 2015
WASHINGTON, D.C.

Deputy Assistant Secretary of the Air Force for Science, Technology, and Engineering
    Dr. David Walker (SES)

Air Force Research Laboratory
    Maj Gen Thomas Masiello, Commander

Air Force Office of Scientific Research
    Dr. Thomas Christian (SES), Director

Panel Discussion with AFOSR Program Staff on "Finding"
    Dr. John Luginsland, Program Officer, Laser and Optical Physics
    Dr. Enrique Parra, Program Officer, Ultrashort Pulse Laser-Matter Interactions
    Lt Col Victor Putz, Program Officer, EOARD Physics
    Maj Justin Silverman, Staff Judge Advocate

Panel Discussion with AFOSR Program Staff on "Funding"
    Mr. Mark Amundson, Program Element Monitor, Basic Research (AQRT)
    Dr. Michael Berman, Program Officer, Molecular Dynamics and Theoretical Chemistry
    Mr. Phillip Cherbaka, Director, Information Technology and Chief Information Officer
    Mr. Rickey Lawrence, Chief, Finance
    Ms. Dorothy Howe, Deputy Chief of Contracting

Panel Discussion with AFOSR Program Staff on "Forwarding"
    Dr. Van Blackwood, Principal Assistant to the Chief Scientist
    Dr. Tatjana Curcic, Program Officer, Atomic and Molecular Physics

APPENDIX C

    Dr. Hugh Delong, Program Officer, Natural Materials and Systems
    Ms. Molly Lachance, Program Analyst, Digital Outreach
    Dr. Kent Miller, Program Officer, Space Sciences

Former AFOSR Directors
    Dr. Joe Janni
    Dr. Lyle Schwartz
    Dr. Pat Carrick (SES)
    Dr. Tom Russell (SES)
    Dr. Brendan Godfrey

Office of the Assistant Secretary of Defense for Research and Engineering
    Dr. Robbin Staffin (SES), Director of Basic Research

Defense Technical Information Center
    George Lembrick (SES), Deputy Administrator
    Christopher Thomas (SES), Administrator

Office of Science and Technology Policy
    Chris Fall, Assistant Director for Defense Programs

Chief Scientists from the Air Force Major Commands
    Janet Fender, Air Combat Command
    Don Erbschloe, Air Mobility Command
    Doug Beason, Air Force Space Command (retired)
    Azar Ali, Pacific Air Forces

SESSION 2
MAY 27-28, 2015
WASHINGTON, D.C.

Air Force Chief Scientist
    Dr. Mica Endsley

Air Force Research Laboratory
    Dr. Morley Stone, Chief Technology Officer

Past NSF Director's Perspective
    Dr. Rita Colwell, Distinguished University Professor, University of Maryland

Army Research Office
    Dr. Dave Skatrud (SES), Director

Department of Energy Office of Science
    Dr. Linda Blevins, Senior Technical Advisor, Office of the Deputy Director for Science Programs

Office of Naval Research
    Dr. Walter Jones (SES), Executive Director

University Approaches to Managing Basic Research
    Dr. Bill Bonvillian, Director, MIT Washington Office
    Dr. Steve Cross, Executive Vice President for Research, Georgia Institute of Technology
    Dr. Denis Wirtz, Vice Provost for Research, Johns Hopkins University